The LogicPrep Guide to Geometry

LogicPrep

About this book

This book is the culmination of many hours of diligent work by the following people: Jesse Kolber, Roger Reiersen, Jamie Kenyon, Helen Moss, Brian Siberine, Molly Pickel, Matthew Kwong, Chad Schwam, and Alex Wurm. Our dynamic team collectively has over 30 years in the test-prep industry and 20,000 hours of preparing students for both the ACT and SAT.

Published by LogicPrep Tutoring, 2016
ISBN: 978-0-9851060-4-1

Copyright Notice
© LogicPrep Tutoring. All Rights Reserved.
Copying and electronic transmission are strictly forbidden, including, but not limited to, e-mail, facsimile, CD-ROM, DVD-ROM, tape, or any medium for the purpose of distribution. Printing and distribution of this publication is strictly prohibited without the authors' expressed written permission.

Contents

A Note from LogicPrep	1
Number Lines and Plotting Points	3
Midpoint and Distance Drill	11
Midpoint and Distance	14
Line Between Two Points Drill	21
Lines and Slopes	23
Parallel and Perpendicular Lines Drill	35
Parallel and Perpendicular Lines	37
Lines and Intercepts Drill	45
Lines and Intercepts	47
Scatterplots	51
Reflections and Transformations	55
Inequalities	64
Absolute Value	70
Functions	80
Function Shifts	87
Quadratics	94
Circle Formula	101
Intersections	106
Logic	110
Coordinate Geometry Mixed Problem Set	115
Parallel Lines and Transversals	140
Lines and Angles	147
Pythagorean Theorem	160
Triangle Inequality Theorem	164
Similar Triangles	167
Isosceles Triangles	174
Triangle Area	178
Special Right Triangle Drill	187
Squares and 45-45-90 Triangles	191
Equilateral and 30-60-90 Triangles	195
Polygons	200
Parallelograms	205
Trapezoids	211
Circles I Drill	221
Circles	223
Circles II Drill	233
Inscribed Angles	237
Circles III Drill	243
Circle Sectors	246

Contents

Shaded Region 251
Plane Geometry Mixed Problem Set 257
Surface Area 290
Volume 295
Diagonals 302
Ratios and Dimensions 307
Logical Spatial Relationships 309
Intro to Trigonometry 315
Formulas Drill 329

A Note from LogicPrep

Hi there. Welcome to the high-stakes, much-dreaded, your-whole-life-seems-to-hang-in-the-balance world of college entry exams.

Here's the good news:

If you're reading this book, you're already smart – not because you chose our book (well, not *just* because), but because you've realized that the best way to get the ACT or SAT score that you want is to prepare. Prepare logically, even.

To be sure, your ACT or SAT score will be only a part of your entire college application. Your grades, extracurriculars, outside activities, volunteer work, and of course your main essay and supplements will all factor in. But let's be honest: your test scores *will* be a big part of your application.

How did this come to be, you ask? Why are there standardized tests like the ACT and SAT in the first place? There are many reasons, but the biggest one is basically that the American education system is set up on a district-by-district basis – so colleges need a uniform national standard to find out how much you know and where you fit in with your peers. Thus, the ACT and the SAT test how much you know (of what the test-makers deem important, that is) and where your score fits in on a national bell curve. Simple, right?

This book is organized to teach you clearly and thoroughly the primary geometry skills that you'll need on the math sections of the ACT and SAT. After you master these skills, you'll need to learn how the ACT and SAT expect you to apply them. And, even more importantly, you'll need – through thorough reflection and rigorous reflection – to learn how you best take the test.

We're not in the testing room with you, but we will offer you our wisdom, honed from years of working with all different kinds of students, so we can alert you to common (and sometimes not-so-common) mistakes… and some variations on test-taking strategy so that you can find what works best for you.

Finally, after you learn all the concepts the ACT and SAT test and identify the test strategies that work best for you, there are three more skills (which may not sound like skills) that you'll need to master: being confident, being calm, and being careful.

- **BE CONFIDENT** – You have learned the knowledge you need and have seen how the ACT and SAT test it, so be strong! Neither the ACT nor the SAT varies much from test to test.
- **BE CALM** – Becoming emotional on these tests will work against you. Whether you feel you are performing well or not, you must stay calm instead of becoming excited or nervous.
- **BE CAREFUL** – Even experienced test-takers can be sloppy, or can misread a question, or can forget to double check exactly what the question is asking for… and so they let themselves get caught off-guard by a well-written but incorrect answer choice.

The road to success will be a little different for everyone, and the exact final destinations will vary, too. But here's the promise we make:

If you work hard to learn the skills we teach, and if you practice and adjust your test-taking skills, your score will improve – often by a lot. How much improvement you attain depends on *you*: on *your* hard work and on *your* commitment.

Lastly, we do teach these skill sets so you can do great on the ACT and SAT, but we wouldn't be in this business if we didn't also believe we were teaching you things that would be vital for the rest of your life.

So jump in. The water's fine. Have some fun and know you are making an investment that will pay off – on the ACT and SAT, and beyond!

– The LogicPrep Team

Number Lines and Plotting Points
Geometry Problem Set 1

1. On the number line below, the tick marks are equally spaced and as shown. Of these coordinates, which has the greatest negative value?

   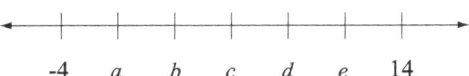

   A. a
   B. b
   C. c
   D. d
   E. e

2. Points P, Q, R, and S lie on a line, in that order. If $\overline{PS} = 17$, $\overline{PR} = 11$ and $\overline{QS} = 9$, what is the length of $\overline{QR}$?

   F. 1
   G. 2
   H. 3
   J. 4
   K. 5

3. On the line below, if the lines segments $\overline{NO}$, $\overline{LO}$ and $\overline{KN}$ have lengths of 3, 10, and 13, respectively, what is the length of segment $\overline{KL}$?

   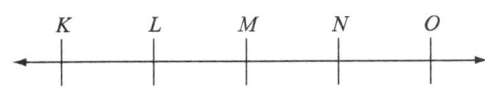

   Figure not drawn to scale

   A. 3
   B. 6
   C. 7
   D. 8
   E. 13

Number Lines and Plotting Points
Geometry Problem Set 1

4. If the tick marks on the number line below are equally spaced, which of the following points, J through N, is between $\frac{1}{4}$ and $\frac{5}{8}$?

   F. J
   G. K
   H. L
   J. M
   K. N

5. If the length of $\overline{AB}$ is 4 and the length of $\overline{BC}$ is 9, which of the following could be the length of $\overline{AC}$?

   A. 4
   B. 12
   C. 14
   D. 19
   E. 25

Number Lines and Plotting Points
Geometry Problem Set 1

6. Which of the lettered points in the figure below has coordinates (x, y) such that $x + y = 2$?

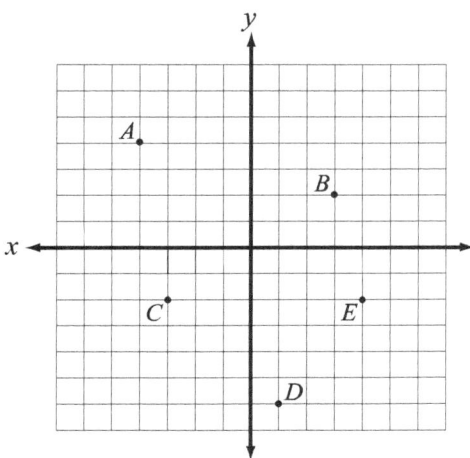

F. A
G. B
H. C
J. D
K. E

7. In the figure below, $\overline{PQ} = \overline{RS}$. What is the value of t?

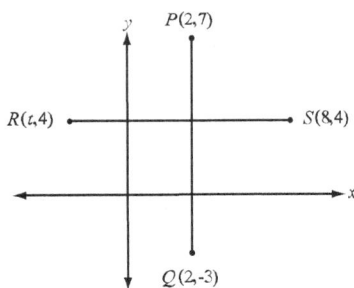

A. -10
B. -4
C. -3
D. -2
E. 2

Number Lines and Plotting Points
Geometry Problem Set 1

8. In the xy-plane below, the circle is tangent to the x-axis and y-axis. What are the coordinates of point A?

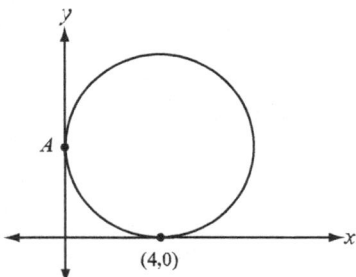

F. $(2, 0)$
G. $(2, 2)$
H. $(0, 2)$
J. $(0, 4)$
K. $(4, 4)$

9. In the figure below, point A is the same distance from the origin as point B is from the origin. If the coordinates of A are (x, y), which of the following must be the coordinates of point B?

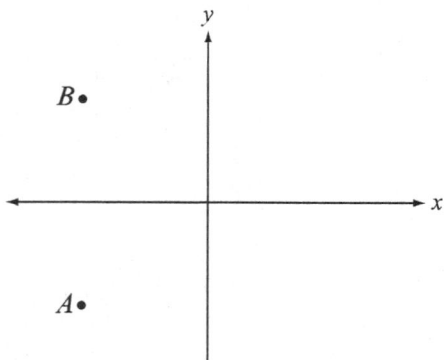

A. (x, y)
B. $(-x, y)$
C. $(x, -y)$
D. $(-x, -y)$
E. (y, x)

Number Lines and Plotting Points
Geometry Problem Set 1

10. On the number line below, the tick marks are equally spaced. What is the value of $b-a$?

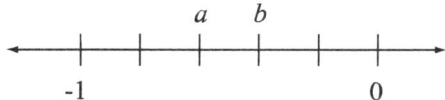

F. -1

G. -.2

H. .2

J. .4

K. 1

11. In the xy-plane, the center of a circle has coordinates $(-1, 3)$. If one endpoint of a diameter of the circle is $(-1, -4)$, what are the coordinates of the other endpoint of this diameter?

A. $(-1, -1)$

B. $(-4, -4)$

C. $(3, -4)$

D. $(-4, 10)$

E. $(-1, 10)$

Number Lines and Plotting Points
Geometry Problem Set 1

12. In the figure below, $\overline{PQ}$ is a diagonal of a square (not shown). Which of the following are the coordinates of one point on the other diagonal of the square?

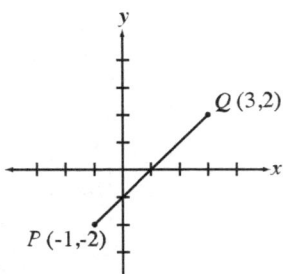

F. $(0, 0)$
G. $(-1, 0)$
H. $(1, -2)$
J. $(-1, 2)$
K. $(0, 2)$

DO YOUR FIGURING HERE

Number Lines and Plotting Points
Geometry Problem Set 1

Answer Key

#	Answer	Frequency	Difficulty
1	A	popular	3
2	H	popular	3
3	B	popular	3
4	H	popular	2
5	B	popular	3
6	K	popular	2
7	D	popular	2
8	J	popular	2
9	C	popular	3
10	H	popular	3
11	E	popular	3
12	J	popular	2

Midpoint and Distance
Quick Drill

1. What is the midpoint of a line from (1, 3) to (1, 9)?

2. A line from (2, y) to (-1, 4) is 5 units long. Find y.

3. What is the midpoint of a line from (6, -2) to (-3, 10)?

4. What is the distance between (3, -4) and (8, -16)?

5. The midpoint of a line from (x, y) to (6, -2) is (4, 1). Find x and y.

6. What is the midpoint of a line from (-1, 4) to (-3, 8)?

7. What is the distance between (-8, 3) and (1, 7)?

8. The midpoint of a line from (3, 4) to (x, y) is (0, -2). Find x and y.

9. What is the distance between (0, -2) and (3, 2)?

10. A line from (0, 4) to (x, 6) is 3 units long. Find x.

Midpoint and Distance
Quick Drill

11. What is the distance from $(9, 4)$ to $(17, 18)$?

12. What is the distance from (a, b) to $(0, b)$?

13. A certain line begins at $(-12, -2)$, ends at $(-7, y)$, and is 13 units long. What is a possible value of y?

14. What is the distance between $(-1, 3)$ and (x, y) if the midpoint of the line is $(2, -1)$?

15. What is the midpoint of a line from $(77, 182)$ to the origin?

16. What is the distance from $\left(\sqrt{3}, 2\right)$ to $\left(3\sqrt{3}, 4\right)$?

17. What is the distance from $\left(130, -2\sqrt{2}\right)$ to $\left(16, 10\sqrt{2}\right)$?

18. A square $ABCD$ has an area of 25 and corners $A(-1, -12)$ and $B(2, y)$. What is a possible value of y?

19. Line $\overline{AB}$ is bisected at $(32, 8)$. The coordinates of A are $(2, 3)$. What are the coordinates of point B?

20. $\overline{AB} = \overline{BC}$ for three distinct points $A(-5, -2)$, $B(7, 3)$, and $C(7, y)$. What is a possible value of y?

Midpoint and Distance
Quick Drill

Answer Key

#	Answer
1	6
2	$y = 0, 8$
3	$\left(\dfrac{3}{2}, 4\right)$
4	13
5	$x = 2$ and $y = 4$
6	$(-2, 6)$
7	$\sqrt{97}$
8	$x = -3$ and $y = -8$
9	5
10	$x = \pm\sqrt{5}$
11	$2\sqrt{65}$
12	a
13	$y = -14, 10$
14	10
15	$(38.5, 91)$
16	4
17	$18\sqrt{41}$
18	$y = -8, -16$
19	$(62, 13)$
20	$y = -10, 16$

Midpoint and Distance
Geometry Problem Set 2

1. Point R is the midpoint of $\overline{QS}$ in the figure below. What is the value of a?

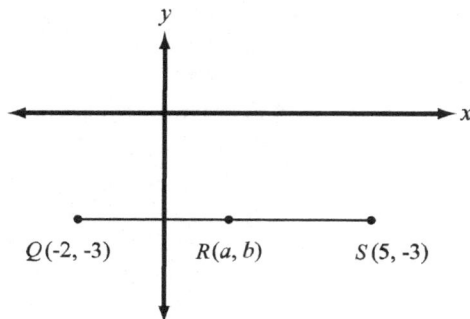

   A. .5
   B. 1
   C. 1.5
   D. 2
   E. 2.5

2. In the xy-coordinate plane below, the distance between points J and K is 9. What are the coordinates of point K?

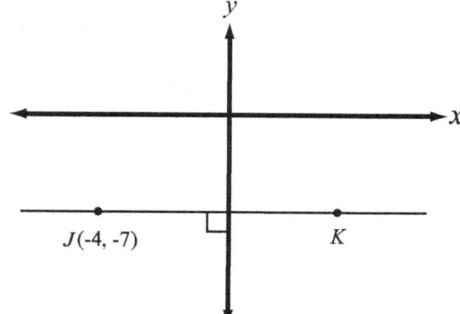

   F. (-4, 2)
   G. (-5, 7)
   H. (5, -7)
   J. (5, 7)
   K. (9, -7)

Midpoint and Distance
Geometry Problem Set 2

3. In the xy-coordinate plane, the distance from point A to point $(2, 5)$ is 7. If the x-coordinate of A is 2, which of the following could be the y-coordinate of A?

   A. -5
   B. -4
   C. -3
   D. -2
   E. -1

4. On a real number line, point A has coordinate -9 and point B has coordinate -13. What is the coordinate of the midpoint of $\overline{AB}$?

   F. -11
   G. -4
   H. -1
   J. 4
   K. 11

5. In the standard xy-coordinate plane, what is the midpoint of the line segment that has endpoints $(-4, 3)$ and $(1, -1)$?

   A. $(-\frac{3}{2}, 1)$
   B. $(-\frac{5}{2}, 1)$
   C. $(\frac{5}{2}, -1)$
   D. $(3, 1)$
   E. $(3, 2)$

6. In the xy-coordinate plane, the distance from point $(x, 3)$ to point $(3, 7)$ is 5. What is a possible value of x?

   F. -6
   G. -1
   H. 0
   J. 2
   K. 4

Midpoint and Distance
Geometry Problem Set 2

7. In the xy-coordinate plane, point $P(0,-2)$ and point $R(8,6)$ are opposite vertices of square $PQRS$. If point A is the midpoint of line $\overline{PS}$, then what is the area of $\triangle PAQ$?

    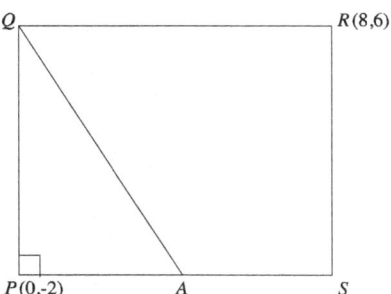

    A. 16
    B. 21
    C. 28
    D. 32
    E. 64

8. In the xy-coordinate plane, segment $\overline{AB}$ has a midpoint M. If A has the coordinates $(1,8)$ and B has the coordinates $(1,72)$, then what is the length of segment $\overline{AM}$?

    F. 4
    G. 32
    H. 40
    J. 64
    K. 80

9. In the xy-coordinate plane, points $A, B, C,$ and D lie on a line in that order. B is the midpoint of line $\overline{AD}$. If the length of $\overline{AD}$ is 12 and the length of line $\overline{BC}$ is 2, then what is the length of line $\overline{CD}$?

    A. 2
    B. 4
    C. 5
    D. 6
    E. 10

Midpoint and Distance
Geometry Problem Set 2

10. In the standard xy-coordinate plane, the midpoint between $(-10, 8)$ and which point is $(-2, 3)$?

    F. $(-6, 5.5)$

    G. $(-6, 2)$

    H. $(6, -2)$

    J. $(6, 4)$

    K. $(8, -5)$

11. In the xy-coordinate plane, point A is $(-3, 2)$, point B is $(2, 2)$ and point C is $(-3, -4)$. What is the perimeter of the triangle formed by connecting A, B and C?

    A. 15

    B. 22

    C. $11 + \sqrt{30}$

    D. $11 + \sqrt{61}$

    E. $\sqrt{82}$

12. In the xy-coordinate plane, point $Q(4, 8)$ and point $S(-6, -2)$ are opposite vertices of square $PQRS$. If point A is the midpoint of line $\overline{SR}$, what is the area of $\triangle PAQ$?

13. In the xy-coordinate plane, the distance between points $A(27, y)$ and $B(12, 15)$ is 17. What is one possible value of y?

Midpoint and Distance
Geometry Problem Set 2

Answer Key

#	Answer	Frequency	Difficulty
1	C	popular	2
2	H	popular	2
3	D	popular	2
4	F	popular	1
5	A	popular	2
6	H	popular	2
7	A	popular	2
8	G	popular	1
9	B	popular	1
10	F	popular	2
11	D	popular	3
12	50	popular	4
13	7 or 23	popular	4

Line Between Two Points
Quick Drill

1. What is the slope of a line from $(2, 8)$ to $(-3, 6)$?

2. What is the equation in slope-intercept form for a line from $(-4, 2)$ to $(3, 0)$?

3. A line has a slope of $-\frac{1}{3}$ and passes through $(17, 8)$ and $(4, y)$. What is the value of y?

4. What is the slope of a line from $(0, a)$ to (a, a)?

5. A line with the equation $\frac{14y}{3} = 6x - \frac{18}{5}$ passes through $(x, 16)$. What is the value of x?

6. A line passes through the origin and $(2, -5)$. What is the equation of the line in slope-intercept form?

7. What is the equation in slope-intercept form for a line from $(0, 22)$ to $(-18, 6)$?

8. Give the equation of a line from $(12, -2)$ to $(13, 8)$ in point-slope form.

9. Give the equation of a line from $(2, 3)$ to $(5, 8)$ in point-slope form.

10. Give the equation of a line from $(0, 4)$ to $(-2, -8)$ in point-slope form.

Line Between Two Points
Quick Drill

Answer Key

#	Answer
1	$\frac{2}{5}$
2	$y = -\frac{2x+6}{7}$
3	$y = \frac{37}{3}$ or $12\frac{1}{3}$
4	0
5	$x = \frac{587}{45}$
6	$y = -\frac{5}{2}x$
7	$y = -\frac{8}{9}x + 22$
8	$y + 2 = 10(x - $ or $y - 8 = 10(x - 13)$
9	$y - 3 = \frac{5}{3}(x - 2)$ or $y - 8 = \frac{5}{3}(x - 5)$
10	$y - 4 = 6(x - 0)$ or $y + 8 = 6(x + 2)$

Lines and Slopes
Geometry Problem Set 3

1. In the standard xy-coordinate plane, what is the slope of the line through $(6, 9)$ and $(2, -5)$?

    A. $-\dfrac{7}{2}$

    B. $-\dfrac{2}{7}$

    C. 1

    D. $\dfrac{2}{7}$

    E. $\dfrac{7}{2}$

2. In the figure below, line m intersects the x-axis at $x = 1$ and the y-axis at $y = 3$. If line n (not shown) is perpendicular to line m, what is the slope of line n?

    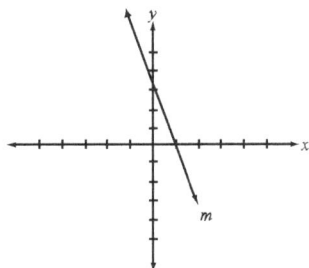

3. Point A lies on the line described by the equation $y + 3 = (x - 8)$. If the y-coordinate of A is 5, what is the x-coordinate of A?

4. In the xy-coordinate plane, the line $3x - y = a$ passes through the point $(8, 15)$. What is the value of a?

Lines and Slopes
Geometry Problem Set 3

5. Which of the following linear functions has a slope equal to -3?

A.
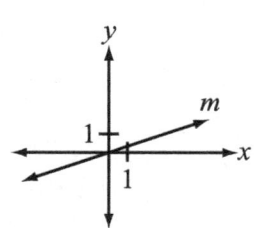

B.
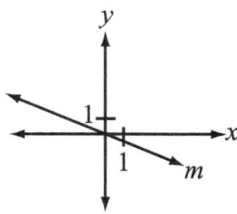

C.
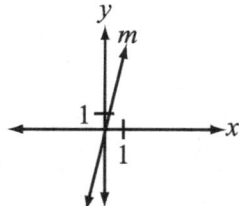

D.
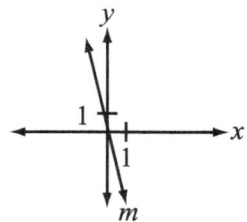

E.
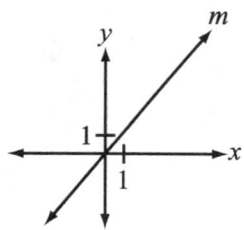

DO YOUR FIGURING HERE

6. Which of the following lines contains the point (-3, 7)?

F. $3x + y = 2$

G. $x + 3y = -2$

H. $4x + y = 5$

J. $x + 3y = 5$

K. $3x + 2y = 5$

Lines and Slopes
Geometry Problem Set 3

7. A line contains the points $(1,-2)$ and $(3,6)$. Which of the following could be the equation for this line?

   A. $4x - y = 6$

   B. $x - 4y = 4$

   C. $4x + y = 8$

   D. $x + 4y = 12$

   E. $12x + 2y = 16$

8. In the figure below, which of the following line segments (not shown) has a slope of $\dfrac{1}{2}$?

   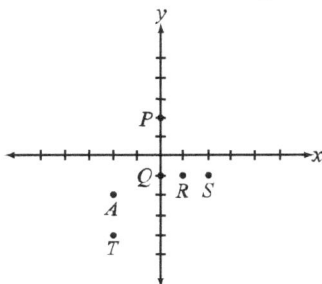

   F. $\overline{AP}$

   G. $\overline{AQ}$

   H. $\overline{AR}$

   J. $\overline{AS}$

   K. $\overline{AT}$

Lines and Slopes
Geometry Problem Set 3

9. In the figure below, what is the slope of line l?

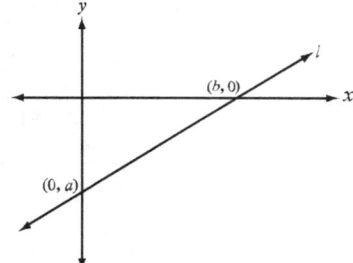

- A. $\dfrac{a}{b}$
- B. $\dfrac{-a}{b}$
- C. $\dfrac{b}{a}$
- D. $\dfrac{-b}{a}$
- E. a

10. What is the slope of line m in the figure below?

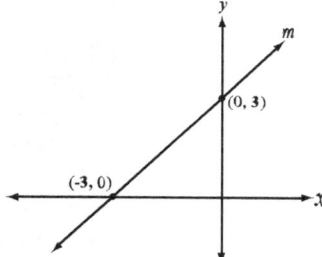

- F. -3
- G. -1
- H. $\dfrac{1}{3}$
- J. 1
- K. 3

Lines and Slopes
Geometry Problem Set 3

11. In the xy-coordinate plane below, the point $(a, 15)$ (not shown) lies on the line m, which passes through the origin. What is the value of a?

    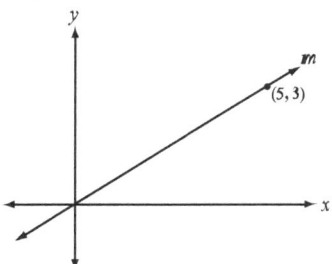

    A. 9
    B. 15
    C. 20
    D. 25
    E. 30

12. In the xy-coordinate plane, $y = 4x + 7$ and $y = mx + b$ are parallel lines. What is the value of m?

    F. $\dfrac{-1}{4}$
    G. $\dfrac{1}{7}$
    H. $\dfrac{1}{4}$
    J. 4
    K. 7

13. Line m lies in the xy-coordinate plane and contains points $(2, 9)$ and $(-1, 1)$. If line n is perpendicular to line m, what is the slope of line n?

    A. $-\dfrac{8}{3}$
    B. $-\dfrac{3}{8}$
    C. $-\dfrac{1}{3}$
    D. $\dfrac{3}{8}$
    E. $\dfrac{8}{3}$

Lines and Slopes
Geometry Problem Set 3

14. In the xy-coordinate plane, the line $y = mx + 7$ is perpendicular to the line $4x + 7y = 3$. What is the value of m?

- F. $-\dfrac{7}{4}$
- G. $-\dfrac{4}{7}$
- H. $-\dfrac{1}{4}$
- J. $\dfrac{4}{7}$
- K. $\dfrac{7}{4}$

15. In the standard xy-coordinate plane, what is the slope of the line joining the points (-5, -9) and (3, -5)?

- A. -2
- B. $-\dfrac{1}{2}$
- C. $\dfrac{1}{2}$
- D. $\dfrac{4}{7}$
- E. $\dfrac{7}{4}$

16. What is the slope of the line given by the equation $6y - 5x = -2$ in the standard xy-coordinate plane?

- F. $-\dfrac{6}{5}$
- G. $-\dfrac{5}{6}$
- H. $\dfrac{5}{6}$
- J. $\dfrac{6}{5}$
- K. 5

DO YOUR FIGURING HERE

Lines and Slopes
Geometry Problem Set 3

17. Line n (not shown) passes through the origin and intersects $\overline{PQ}$ between P and Q. What is one possible value of the slope of n?

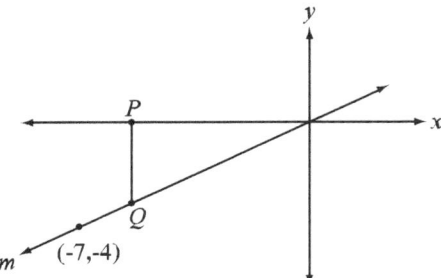

18. In the standard xy-coordinate plane, what is the slope of a line that is perpendicular to $5x + 2y = 14$?

   F. $-\dfrac{5}{2}$

   G. $-\dfrac{2}{5}$

   H. $\dfrac{2}{5}$

   J. $\dfrac{5}{2}$

   K. 5

19. The figure below shows line k in the xy-plane. Line m (not shown) has the equation $y = ax + c$, where a and c are constants. If $m \parallel k$, which of the following must be true?

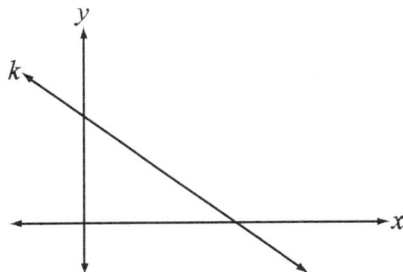

   A. $c < 0$
   B. $c > 0$
   C. $a = 0$
   D. $a < 0$
   E. $a > 0$

Lines and Slopes
Geometry Problem Set 3

20. In the figure below, the slope of line m is $-\frac{1}{2}$. What is the value of a?

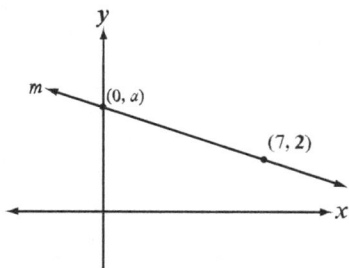

- F. 3
- G. 4
- H. $\frac{11}{2}$
- J. $\frac{-7}{2}$
- K. $\frac{7}{2}$

21. In the coordinate plane below, the coordinates of point P are (a, b), where $|a| > |b|$. Which of the following could be the slope of $\overline{PQ}$?

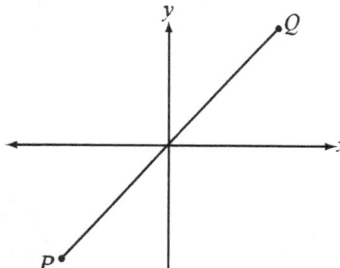

- A. -3
- B. $\frac{-1}{3}$
- C. 0
- D. $\frac{1}{3}$
- E. 3

Lines and Slopes
Geometry Problem Set 3

22. The tables below show selected values for the functions f and g. If both f and g are linear functions, what is the value of $a - b$?

x	$f(x)$	x	$g(x)$
2	3	3	-1
4	a	6	b
6	11	12	8

F. 3.5

G. 5

H. 6

J. 7

K. 11

23. Rectangle $PQRS$ lies in the xy-coordinate plane so that its sides are NOT parallel to the axes. What is the product of the slopes of all four sides of rectangle $PQRS$?

A. -4

B. -1

C. 0

D. 1

E. 4

24.
$$3x + 2y = 11$$
$$9x + ay = 15$$

For which of the following values of a will the system of equations shown above have no solution?

F. -6

G. -4

H. 0

J. 4

K. 6

Lines and Slopes
Geometry Problem Set 3

25. In the xy-coordinate plane below, points A and B are the centers of the circles, which are tangent to the y-axis. What is the slope of line $\overline{AB}$ (not shown)?

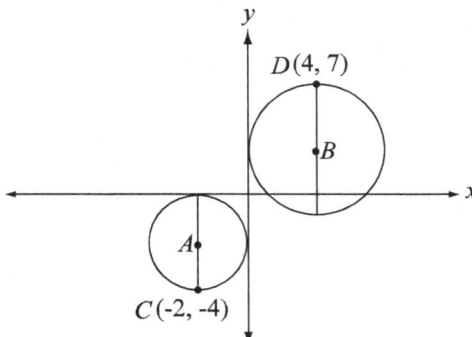

A. $\dfrac{1}{12}$

B. $\dfrac{1}{2}$

C. $\dfrac{5}{6}$

D. $\dfrac{11}{12}$

E. $\dfrac{11}{6}$

DO YOUR FIGURING HERE

Lines and Slopes
Geometry Problem Set 3

Answer Key

#	Answer	Frequency	Difficulty
1	E	popular	2
2	$\frac{1}{3}$ or 0.333	popular	2
3	16	average	1
4	9	popular	2
5	D	popular	1
6	K	average	1
7	A	average	2
8	G	popular	2
9	B	popular	2
10	J	popular	2
11	D	average	2
12	J	popular	1
13	B	popular	2
14	K	popular	2
15	A	popular	1
16	H	popular	2
17	$0 < n < \frac{4}{7}$	popular	2
18	H	popular	1
19	D	popular	3
20	H	popular	3
21	D	popular	3
22	G	average	4
23	D	popular	4
24	K	popular	4
25	C	rare	1

Parallel and Perpendicular Lines
Quick Drill

1. What is the slope of a line that is perpendicular to a line with the equation $y = 3x - 8$

2. What is the slope of a line that is parallel to a line with the equation $y = 6x + 2$?

3. What is the slope of a line that is perpendicular to a line with the equation $y = -\frac{1}{2}x + 3$?

4. What is the slope of a line that is parallel to a line with the equation $4x + 2 = \frac{y+x}{3}$?

5. What is the slope of a line that is perpendicular to a line with the equation $6(x+y) = \frac{13}{2} + 2y$?

6. What is the equation of a line that is parallel to a line with the equation $y = 2x - 1$ and passes through the point $(-2, 1)$?

7. What is the slope of a line that is parallel to a line that passes through $(3, 8)$ and $(-1, 7)$?

8. What is the slope of a line that is perpendicular to a line with the equation $2x - 8y = 3$?

9. What is the equation of a line that is parallel to $\frac{y}{2} = 4x - 3$ and passes through the origin?

10. Give the equation in point-slope form for a line that is perpendicular to $72xy = 2x^2$ and passes through the point $(1, 64)$.

Parallel and Perpendicular Lines
Quick Drill

Answer Key

#	Answer
1	$-\dfrac{1}{3}$
2	6
3	2
4	11
5	$\dfrac{2}{3}$
6	$y = 2x + 5$
7	$\dfrac{1}{4}$
8	-4
9	$y = 8x$
10	$y - 64 = -36(x - 1)$

Parallel and Perpendicular Lines
Geometry Problem Set 4

1. In the figure below, line A is parallel to line B. What is the value of m?

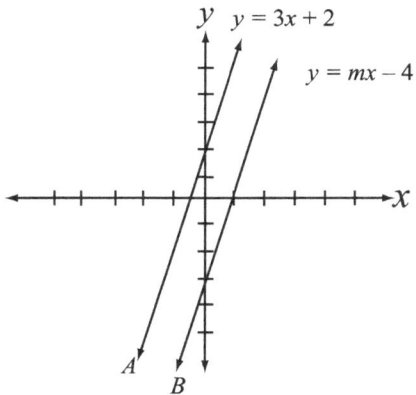

2. In the figure below, lines i and j are parallel. Line k and line l are perpendicular. What is the value of x?

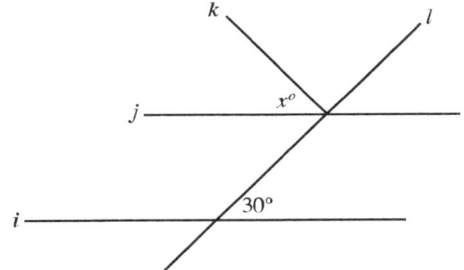

   F. 30
   G. 45
   H. 60
   J. 85
   K. 150

Parallel and Perpendicular Lines
Geometry Problem Set 4

3. In the xy-coordinate plane, $y = 4x + 7$ and $y = mx + b$ are parallel lines. What is the value of m?

   A. $-\dfrac{1}{4}$

   B. $\dfrac{1}{7}$

   C. $\dfrac{1}{4}$

   D. 4

   E. 7

4. For the two intersecting lines below, which of the following must be true?

   I. $b = 2a$
   II. $a > c$
   III. $c + 110 = a + b$

   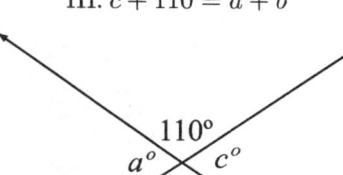

   F. II only

   G. III only

   H. II and III only

   J. I and III only

   K. I, II, and III

DO YOUR FIGURING HERE

Parallel and Perpendicular Lines
Geometry Problem Set 4

5. In the figure below, three lines intersect as shown. What is the value of a?

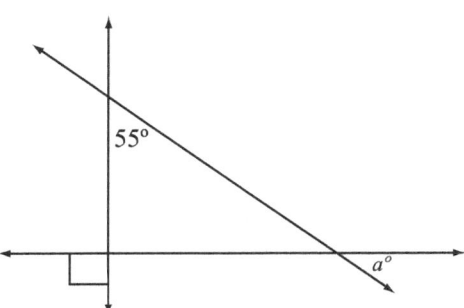

   A. 30
   B. 35
   C. 40
   D. 45
   E. 50

6. In the figure below, line m is parallel to line n. Which of the following angles must be congruent to $\angle x$?

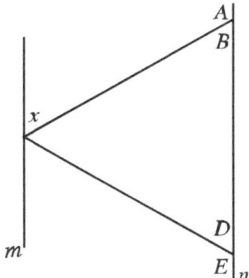

   F. $\angle A$
   G. $\angle B$
   H. $\angle C$
   J. $\angle D$
   K. $\angle E$

Parallel and Perpendicular Lines
Geometry Problem Set 4

7. The slope of line l in the standard xy-coordinate plane is $\frac{3}{5}$. Which of the following is an equation of a line that is perpendicular to line l?

   A. $5x - 3y = 6$
   B. $5x + 3y = 6$
   C. $3x + 5y = 6$
   D. $3x - 5y = 6$
   E. $-3x + 5y = 6$

8. In the xy-coordinate plane, line m has the equation $6 = -24x + 6y$. Line n is parallel to line m and passes through the point $(3, 2)$. What is the equation of line n?

   F. $y = -24x + 74$
   G. $y = 4x - 5$
   H. $y = 4x - 10$
   J. $y = 6x$
   K. $y = 8x - 5$

9. In the xy-coordinate plane, line $y = mx + 142$ is perpendicular to line $8y - 14x = 16$. What is the value of m?

   A. -7
   B. $-\frac{4}{7}$
   C. $\frac{1}{16}$
   D. $\frac{7}{4}$
   E. 4

DO YOUR FIGURING HERE

Parallel and Perpendicular Lines
Geometry Problem Set 4

10. Which of the following are equations for lines with a slope of $\frac{1}{4}$?

    I. $x = 4y + 8$
    II. $y = x^2 + x + 1$
    III. $4y = x - 13$

    F. I only
    G. II only
    H. III only
    J. I and III only
    K. I, II, and III

11. Line $\overline{AB}$ is 13 units long. Point A has the coordinates $(-2, 0)$ and point B has the coordinates $(3, y)$. What is a possible value of y?

    A. -12
    B. -6
    C. -2
    D. 4
    E. 5

12. Line l ($4y = 2x + 3$) is reflected over the x-axis to form line m. What is the equation for line m?

    F. $y = -\frac{1}{2}x - \frac{3}{4}$
    G. $y = -2x - \frac{3}{4}$
    H. $y = \frac{1}{2}x + \frac{3}{4}$
    J. $y = 3x + \frac{1}{2}$
    K. $y = x + 6$

Parallel and Perpendicular Lines
Geometry Problem Set 4

13. Line m (not pictured) is perpendicular to line l in the picture below. What is one possible equation for line m?

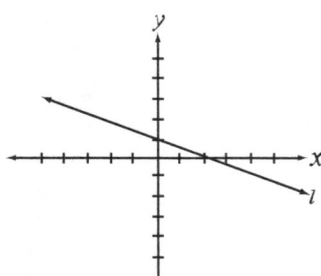

   A. $y = 3x + 2$
   B. $y = x + 1$
   C. $y = 2x - 10$
   D. $y = 4x + 1$
   E. $y = \frac{1}{2}x - 1$

14. In the xy-coordinate plane, the line $y = mx + 3$ is perpendicular to the line $4x + 7y = 3$. What is the value of m?

   F. $-\frac{7}{4}$
   G. $-\frac{4}{7}$
   H. $-\frac{1}{4}$
   J. $\frac{4}{7}$
   K. $\frac{7}{4}$

15. A parallelogram is graphed in the xy-coordinate plane, with verticies $A(-2, 3)$, $B(4, 8)$, and $C(6, 0)$. Which of the following are possible coordinates for the fourth vertex, point D?

   A. $(-2, 0)$
   B. $(0, -5)$
   C. $(1, -3)$
   D. $(2, -6)$
   E. Cannot be determined

Parallel and Perpendicular Lines
Geometry Problem Set 4

Answer Key

#	Answer	Frequency	Difficulty
1	3	popular	1
2	H	average	1
3	D	popular	1
4	G	popular	1
5	B	popular	2
6	G	popular	1
7	B	popular	2
8	H	popular	2
9	B	popular	2
10	J	popular	2
11	A	popular	2
12	F	average	3
13	C	popular	3
14	K	popular	3
15	B	popular	3

Lines and Intercepts
Quick Drill

1. What is the y-intercept of a line with the equation $\frac{3}{2}(x-8) = 6-y$?

2. What is the equation of a line in slope-intercept form that passes through the points $(2,-4)$ and $(7,6)$?

3. What is the slope of a line with a y-intercept of $\left(0, \frac{3}{2}\right)$ that passes through the point $(-2,-5)$?

4. What is the y-intercept of a line with a slope of $-\frac{1}{8}$ and that passes through the point $(8,5)$?

5. What is the equation in slope-intercept form for a line that intercepts the x-axis at 6 and the y-axis at 2?

6. What is the x-intercept of the equation $y^3 + 3y^2 - 4x = 3(x-3) + \frac{y}{x}$?

7. What is the x-intercept of a line with the points $(-2,-14)$ and $(4,-2)$?

8. What is the y-intercept of a line with a slope of -2 and an x-intercept of $(3,0)$?

9. What is the x-intercept of a line with a slope of $\frac{1}{4}$ and a y-intercept of $(0,4)$?

10. What is the x-intercept of a line with the equation $y = \frac{x}{4} - \frac{3}{7}$?

Lines and Intercepts
Quick Drill

Answer Key

#	Answer
1	$(0, 18)$
2	$y = 2x - 8$
3	$\dfrac{13}{4}$ or 3.25
4	$(0, 6)$
5	$y = -\dfrac{1}{3}x + 2$
6	$\left(\dfrac{9}{7}, 0\right)$
7	$(5, 0)$
8	$(0, 6)$
9	$(-16, 0)$
10	$\left(\dfrac{12}{7}, 0\right)$

Lines and Intercepts
Geometry Problem Set 5

1. Which of the following shares an x-intercept with $4y + 2x = 3$?

   A. $4y + x = 3$

   B. $x - 3 = 2y$

   C. $y = \frac{1}{2}x + \frac{3}{2}$

   D. $x = y + \frac{3}{2}$

   E. $2x = 4$

2. The function $y = x^2 + 2x - 3$ intercepts the x-axis at what point(s)?

   F. $(3, 0)$

   G. $(0, 1)$

   H. $(-3, 0)$ and $(1, 0)$

   J. $(0, -3)$ and $(0, 1)$

   K. $(-5, 0)$ and $(5, 0)$

3. In the xy-coordinate plane, the line with equation $y = 3x + 12$ crosses the x-axis at the point with coordinates (p, q). What is the value of p?

   A. -4

   B. 0

   C. 3

   D. 4

   E. 12

4. Line m is determined by the equation $2x + 4y = 12$. At what point does this line cross the y-axis?

   F. $(0, 0)$

   G. $(0, 3)$

   H. $(0, 6)$

   J. $(3, 0)$

   K. $(6, 0)$

DO YOUR FIGURING HERE

Lines and Intercepts
Geometry Problem Set 5

5. Line l is given by the equation $6x + 2y = 18$. Which of the following points is the x-intercept for l?

   A. $(2, 0)$
   B. $(0, 2)$
   C. $(3, 0)$
   D. $(0, 3)$
   E. $(0, 0)$

   DO YOUR FIGURING HERE

6. The graph below shows a parabola whose line of symmetry has equation $x = 2$. If the x-intercepts of the parabola are located at $(n, 0)$ and $(5, 0)$, what is the value of n?

   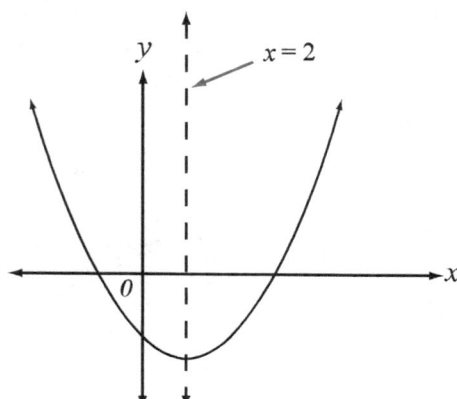

   F. -5
   G. -3
   H. -1
   J. 0
   K. 1

Lines and Intercepts
Geometry Problem Set 5

7. In the figure below, line m has a slope of $-\frac{1}{2}$, and the rectangle under the line has a height of 2 and a width of 6. What is the x-intercept of m?

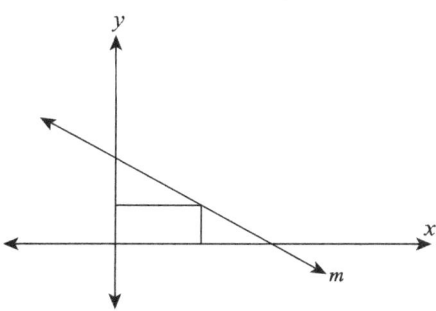

A. 8
B. 9
C. 10
D. 11
E. 12

Lines and Intercepts
Geometry Problem Set 5

Answer Key

#	Answer	Frequency	Difficulty
1	D	average	2
2	H	average	2
3	A	average	2
4	G	average	2
5	C	average	3
6	H	average	3
7	C	popular	3

Scatterplots
Geometry Problem Set 6

1. A student's score on his math tests and the hours he studied for the test are graphed below. If S represents his score and H represents the hours he studied, which of the following equations best describes the data shown?

   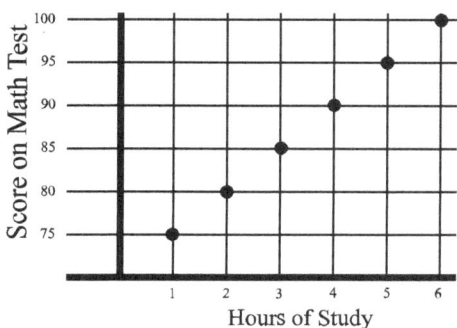

   A. $S = 5H$

   B. $S = 5H + 25$

   C. $S = H + 70$

   D. $S = 5H + 70$

   E. $S = 100 - 5H$

2. Which of the following is true about the line of best fit for the data points graphed below?

   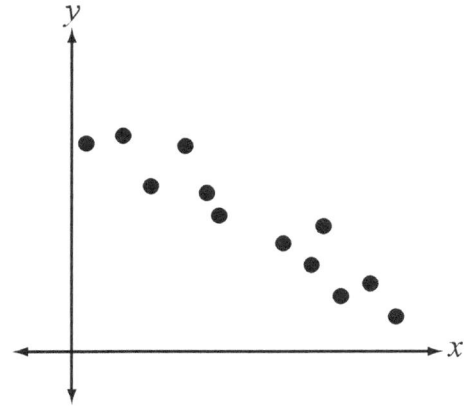

   F. It goes through the point $(0, 0)$

   G. It goes through the point $(0, b)$, where $b < 0$

   H. It goes through the point $(a, 0)$, where $a < 0$

   J. Its slope is negative.

   K. Its slope is positive.

Scatterplots
Geometry Problem Set 6

3. The graph below shows the temperature over a 10 hour period. Which of the following equations best represents the line of best fit for this graph?

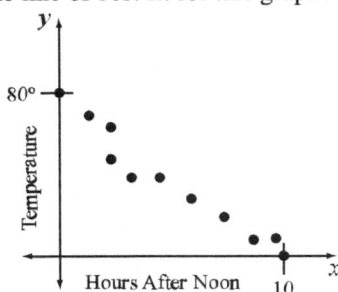

A. $y = 8x + 80$

B. $y = 8x - 80$

C. $y = 8x + 10$

D. $y = 80 - 10x$

E. $y = 80 - 8x$

4. The cost and length of fabric are displayed in the scatterplot below, and the line of best fit for the data is shown. Of the following, which is closest to the average (arithmetic mean) cost, per yard?

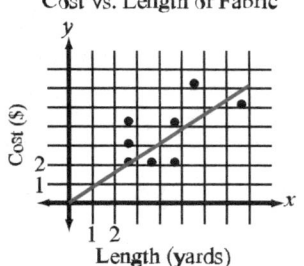

F. $0.50

G. $0.80

H. $1

J. $1.50

K. $2

DO YOUR FIGURING HERE

Scatterplots
Geometry Problem Set 6

5. The graph below shows the amount of milk in a jug after each glass is poured from the jug. Each glass contains x cups of milk. If 6 cups of milk were originally in the jug and $3\frac{3}{4}$ cups remained after each of 9 glasses are poured, what is the value of x?

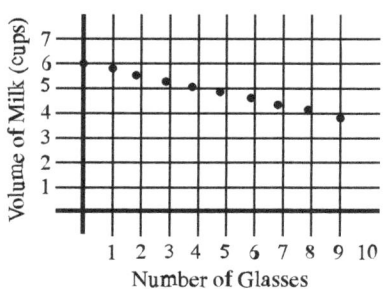

6. The scatterplot below shows the number of female employees and male employees at a company between the years of 1980 and 1988. In which of the years was the percentage of female employees the lowest?

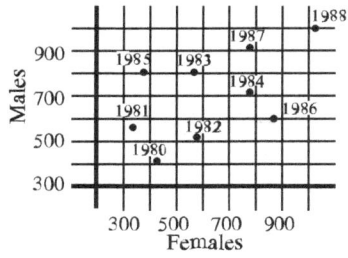

Scatterplots
Geometry Problem Set 6

Answer Key

#	Answer	Frequency	Difficulty
1	D	rare	2
2	J	rare	2
3	E	rare	2
4	G	rare	3
5	.25	rare	3
6	1985	rare	3

Reflections and Transformations
Geometry Problem Set 7

1. Which of the following is the reflection of the graph shown below about the x-axis?

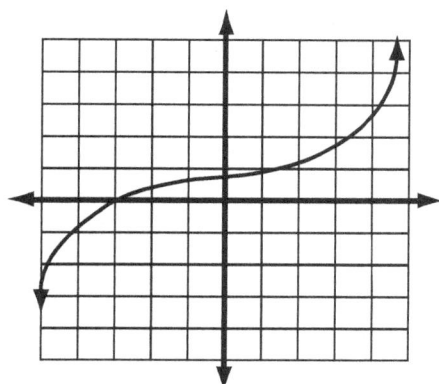

DO YOUR FIGURING HERE

A.

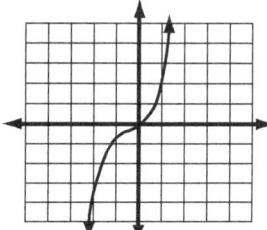

B.

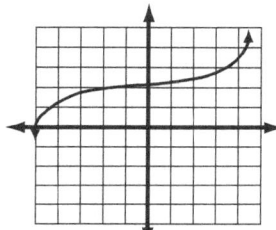

C.

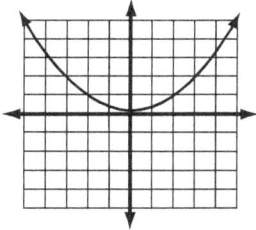

D.

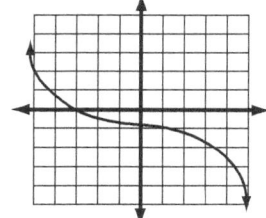

E.

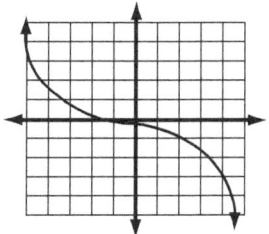

Reflections and Transformations
Geometry Problem Set 7

2. In the xy-plane, line j is formed by reflecting line k around the origin. If the equation of line k is $y = 3x - 2$, what is the slope of line j?

    F. -3

    G. $-\dfrac{1}{3}$

    H. $\dfrac{1}{3}$

    J. 1

    K. 3

3. If the graph below is reflected about the x-axis, what is the total number of x- and y-intercepts that the resulting graph will have?

    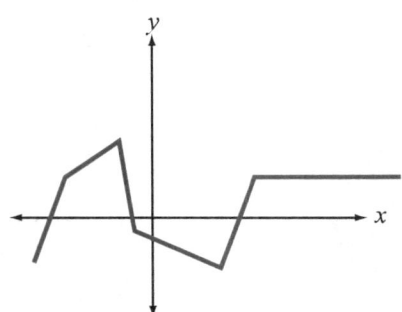

    A. 2

    B. 3

    C. 4

    D. 5

    E. 6

4. In the xy-coordinate plane, three of the vertices of a square are $(0, 0)$, $(-2, 0)$, and $(0, 2)$. If the square is reflected over the line $y = x$, which of the following is a vertex of the reflected square?

    F. $(-2, 2)$

    G. $(2, -2)$

    H. $(0, 2)$

    J. $(2, 2)$

    K. $(-2, 0)$

DO YOUR FIGURING HERE

Reflections and Transformations
Geometry Problem Set 7

5. In the xy-plane, line j is the reflection of line k about the y-axis. If the slope of line j is $\frac{3}{4}$, what is the slope of line k?

    A. $-\frac{4}{3}$

    B. $-\frac{3}{4}$

    C. $-\frac{1}{4}$

    D. $\frac{3}{4}$

    E. $\frac{4}{3}$

6. $\triangle ABC$ is reflected over the line $y = x$. What are the coordinates of C', after the reflection?

    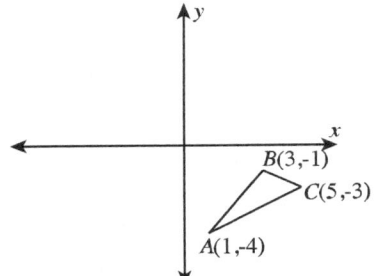

    F. (-5,-3)

    G. (-5, 3)

    H. (-3,-5)

    J. (-3, 5)

    K. (-3,-5)

7. A circle with the center $C(-2, 3)$ is reflected over the line $y = 1$. What is the coordinate of C', after the reflection?

    A. (-2,-3)

    B. (-2,-1)

    C. (-1,-2)

    D. (2,-3)

    E. (2, 3)

Reflections and Transformations
Geometry Problem Set 7

8. $f(x)$ is shown in the xy-coordinate plane below. Which of the following shows $f(x)$ after it has been reflected over the y-axis?

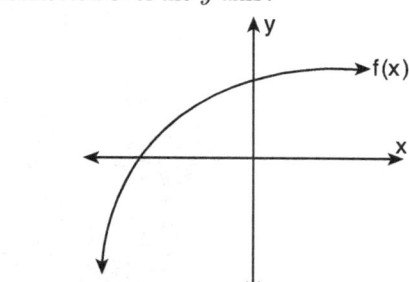

DO YOUR FIGURING HERE

F.
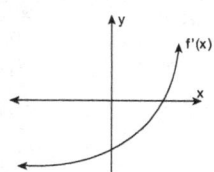

G.
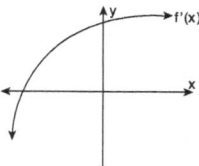

H.
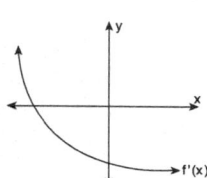

J.
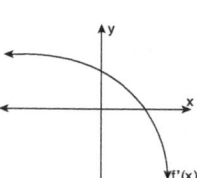

K.
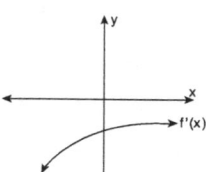

9. Which of the following functions is a reflection of itself over the y-axis?

   A. $f(x) = \sin x$
   B. $f(x) = x$
   C. $f(x) = \dfrac{1}{x}$
   D. $f(x) = x^2$
   E. $f(x) = x^3$

Reflections and Transformations
Geometry Problem Set 7

10. In the xy-plane, the equation of line l is $y = 4x - 7$. If line k is the reflection of line l about the x-axis, what is the equation of line k?

F. $y = -4x - 7$

G. $y = 4x + 7$

H. $y = \frac{1}{4}x + \frac{7}{4}$

J. $y = \frac{1}{4}x - \frac{7}{4}$

K. $y = -4x + 7$

DO YOUR FIGURING HERE

Reflections and Transformations
Geometry Problem Set 7

11. The figure below is to be rotated clockwise 90° about point O. Which of the following shows the resulting figure?

DO YOUR FIGURING HERE

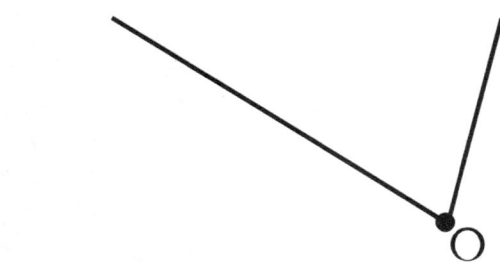

A.

B.

C.

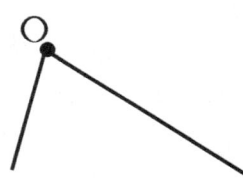

D.

E.

Reflections and Transformations
Geometry Problem Set 7

12. The figure below shows line k in the xy-coordinate plane. Line m (not shown) is obtained by horizontally translating each point on line k 2 units to the left. If the equation of line m is $y = \frac{-3}{4}x + c$, what is the value of c?

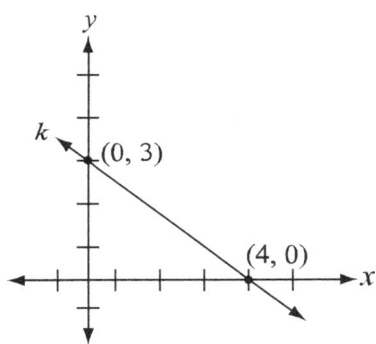

13. In the figure below, $\overline{OA}$ is to be rotated counterclockwise, keeping point O fixed. What will be the coordinates of point A when it first meets the y-axis?

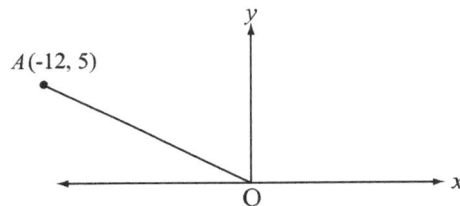

- **A.** $(-12, 0)$
- **B.** $(-13, 0)$
- **C.** $(0, -12)$
- **D.** $(0, -13)$
- **E.** $(0, 13)$

Reflections and Transformations
Geometry Problem Set 7

14. If the figure below were rotated clockwise 270° about point A, which of the following would be the result?

DO YOUR FIGURING HERE

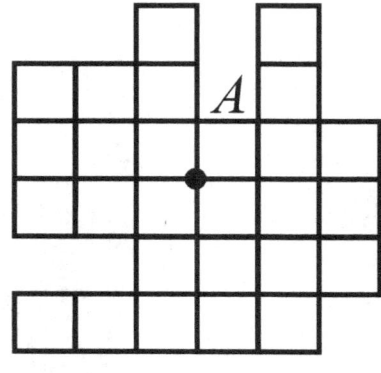

F.

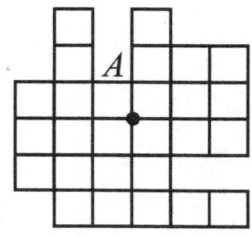

G.

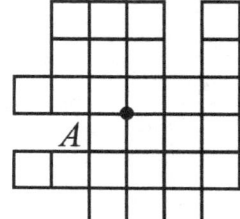

H.

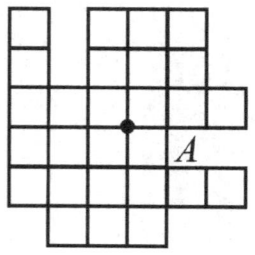

J.

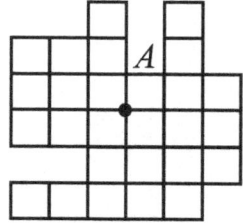

K.

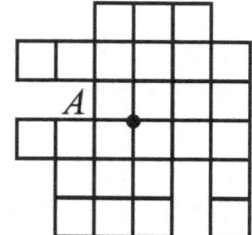

Reflections and Transformations
Geometry Problem Set 7

Answer Key

#	Answer	Frequency	Difficulty
1	D	average	1
2	K	average	2
3	C	average	2
4	G	average	2
5	B	average	2
6	J	average	2
7	B	average	2
8	J	average	1
9	D	average	2
10	K	average	3
11	D	average	1
12	1.5	average	2
13	B	average	2
14	K	average	3

Inequalities
Geometry Problem Set 8

1. If $|3 - 2m| > 16$, which of the following is a possible value for m?

    A. -7
    B. -5
    C. 0
    D. 3
    E. 9

2. What is the solution set of $|2x + 2| \geq 6$?

    F. $\{x : x \geq \text{-}4\}$
    G. $\{x : x \geq 2\}$
    H. $\{x : x \leq \text{-}4 \text{ or } x \geq 2\}$
    J. $x : x \leq \text{-}8 \text{ or } x \geq 4\}$
    K. $\{\ \}$ (the empty set)

3. On a real number line, how far apart are the two solutions to the equation $|x - n| = 9$, for any real number n?

    A. n
    B. $2n$
    C. $n + 9$
    D. $2(n + 9)$
    E. 18

DO YOUR FIGURING HERE

Inequalities
Geometry Problem Set 8

4. Which of the following number line graphs represents the solution set to the inequality $x^2 + 8 \leq 24$?

F.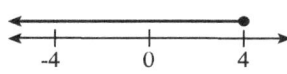

G.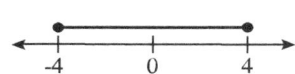

H.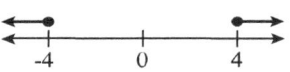

J.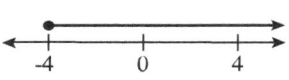

K.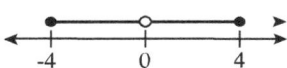

DO YOUR FIGURING HERE

5. Which of the following could represent the graph of $f(x) \geq x - 3$?

A.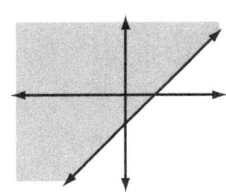

B.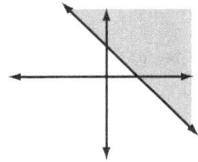

C.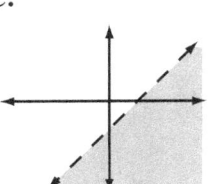

D.

E.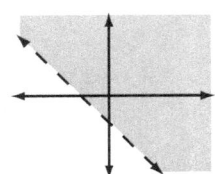

Inequalities
Geometry Problem Set 8

6. Which of the following could represent the graph of $f(x) \leq -x + 1$?

F.
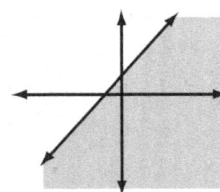

G.
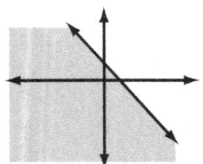

H.
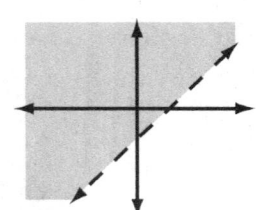

J.
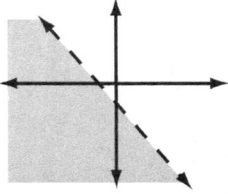

K.
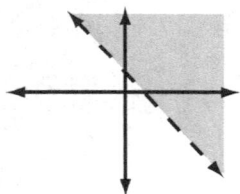

Inequalities
Geometry Problem Set 8

7. The shaded region in the graph below represents the solution set to which of the following sets of inequalities?

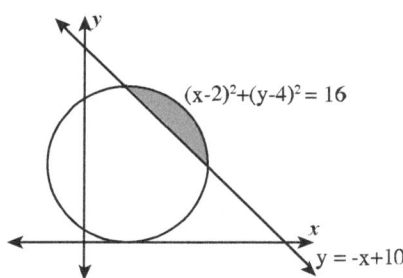

A. $y > -x + 10$ and
$(x - 2)^2 + (y - 4)^2 > 16$

B. $y < -x + 10$ and
$(x - 2)^2 + (y - 4)^2 > 16$

C. $y < -x + 10$ and
$(x - 2)^2 + (y - 4)^2 < 16$

D. $y > -x + 10$ and
$(x - 2)^2 + (y - 4)^2 < 16$

E. $y - 4 < 4$ and $x - 2 > 2$

8. Which of the following shows the solution set to $|x - 8| \leq 8$?

F.
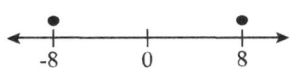

G.

H.
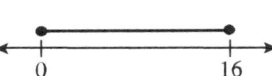

J.
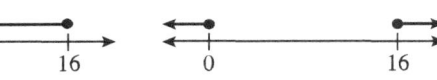

K.
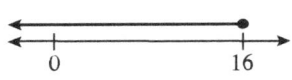

Inequalities
Geometry Problem Set 8

9. Which of the following could show $y > x + 4$ and $y < 3x - 3$?

A.
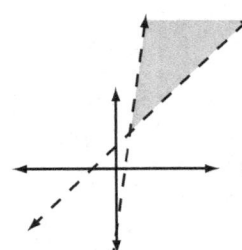

B.
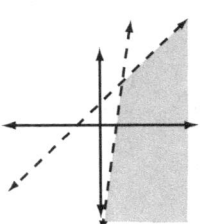

C.
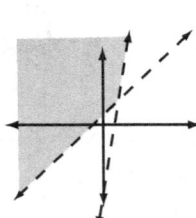

D.
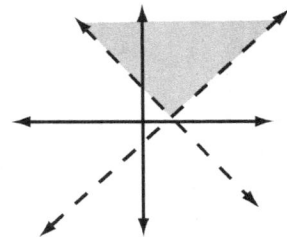

E.
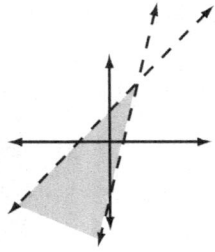

Inequalities
Geometry Problem Set 8

Answer Key

#	Answer	Frequency	Difficulty
1	A	popular	3
2	H	popular	2
3	E	popular	3
4	G	rare	2
5	A	rare	3
6	G	rare	3
7	D	popular	2
8	H	popular	3
9	A	rare	4

Absolute Value
Geometry Problem Set 9

1. Molly turns on the heat if it is 58°F or colder. She turns on a fan if it is 82°F or hotter. Which of the following inequalities shows the temperatures, t, at which neither the heat nor the fan is on?

   A. $t < 82$
   B. $|t - 70| < 12$
   C. $|t - 70| > 24$
   D. $|t - 58| < 24$
   E. $|t - 70| > 12$

2. Which of the points in the xy-plane shown below satisfy the equation $x + |y| = 2$?

   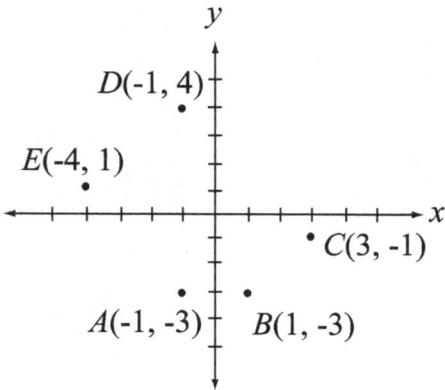

   F. A
   G. B
   H. C
   J. D
   K. E

Absolute Value
Geometry Problem Set 9

3. The figure below shows the graph of $y = f(x)$. Which of the following is the graph of $y = |f(x)|$?

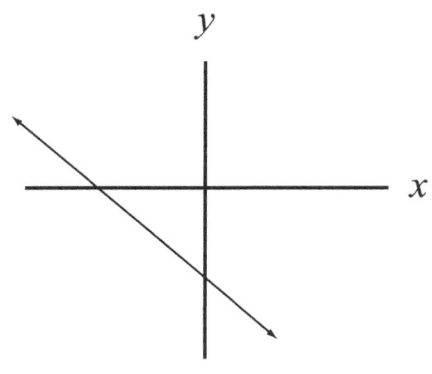

B.

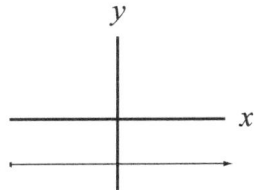

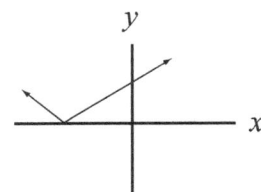

D.

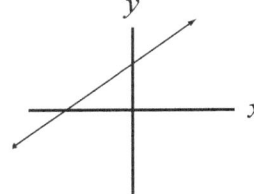

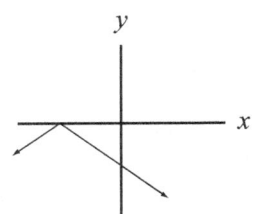

Absolute Value
Geometry Problem Set 9

4. The figure below shows the graph of $y = f(x)$. Which of the following is the graph of $y = -|f(x)|$?

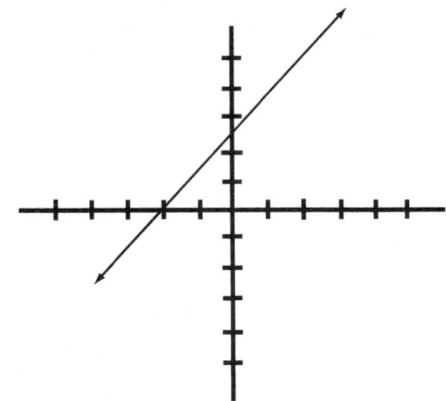

F. G.

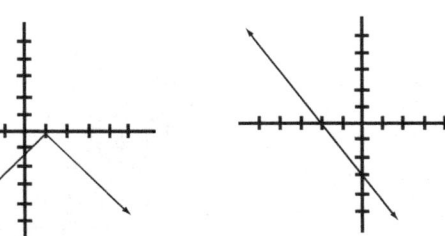

H. J.

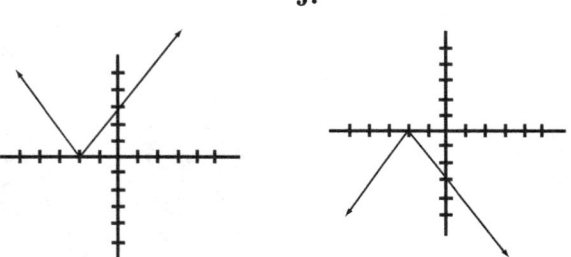

K.

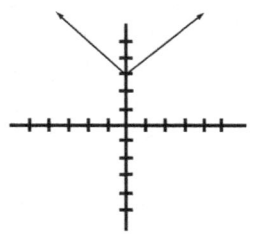

Absolute Value
Geometry Problem Set 9

5. The figure below shows the graph of $y = f(x)$. Which of the following is the graph of $y = |f(x)|$?

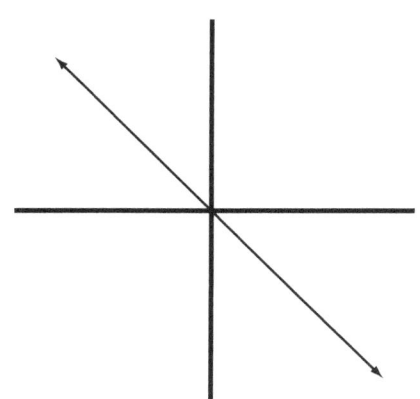

DO YOUR FIGURING HERE

A.   B.

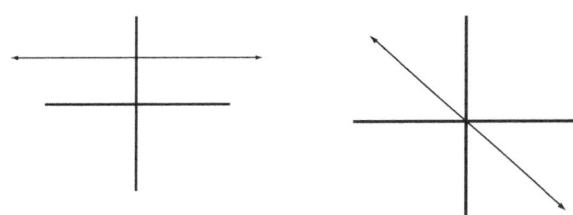

C.   D.

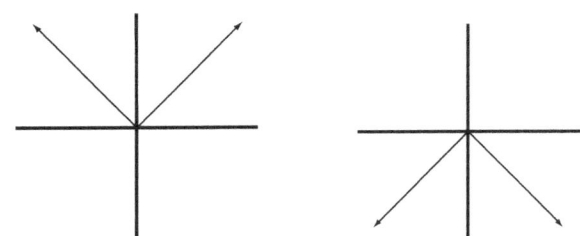

E.

Absolute Value
Geometry Problem Set 9

6. The graph of $y = f(x)$ is shown in the standard (x, y) coordinate plane below. Which of the following graphs is that of $y = |f(x)|$?

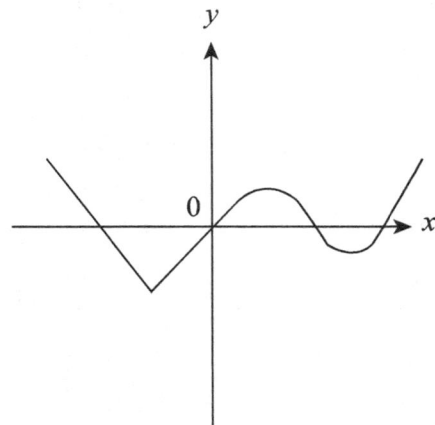

DO YOUR FIGURING HERE

F.
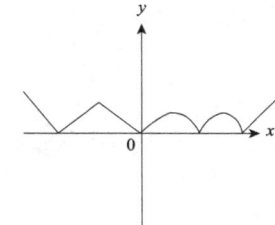

G.

H.
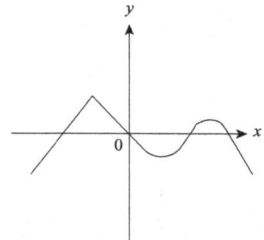

J.
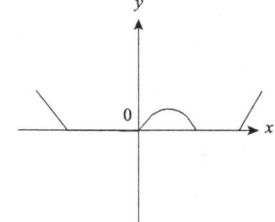

K.
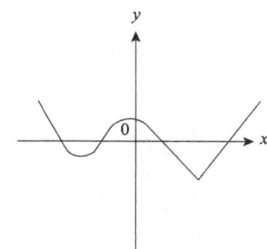

Absolute Value
Geometry Problem Set 9

7. What is the range of the function $g(x) = |x+8|$?
   A. $x > -8$
   B. $x > 0$
   C. $g(x) > -8$
   D. $g(x) > 0$
   E. $g(x) > 8$

8. Which of the following shows the solution set to $|x - 8| \leq 8$?

F.
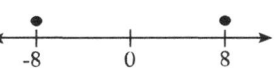

G.
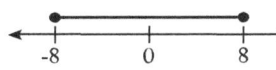

H.
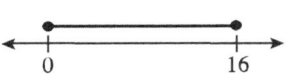

J.

K.
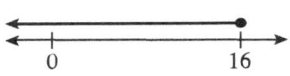

DO YOUR FIGURING HERE

Absolute Value
Geometry Problem Set 9

9. This is the graph of $y = f(x)$
   Which is the graph of $y = |f(x)|$?

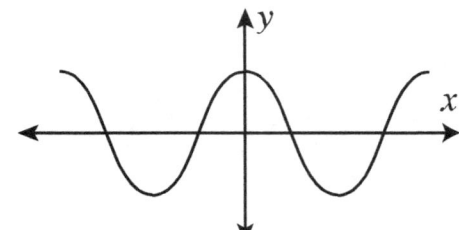

DO YOUR FIGURING HERE

A.

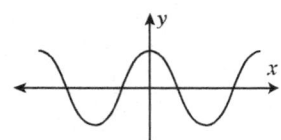

B.

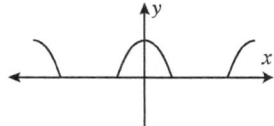

C.

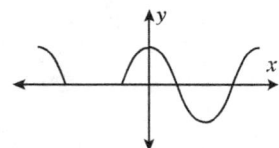

D.

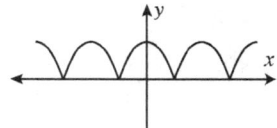

E.

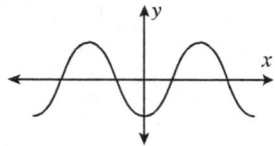

Absolute Value
Geometry Problem Set 9

10. The equation of the line below is $y = \frac{3}{2}x + 3$. Which of the following is the graph of $y = -\left|\frac{3}{2}x + 3\right|$?

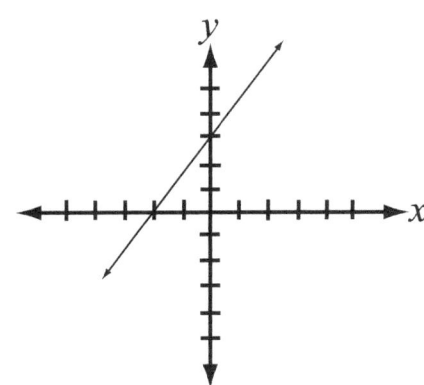

DO YOUR FIGURING HERE

F.
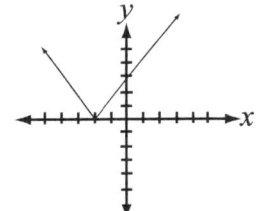

G.
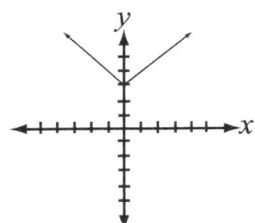

H.
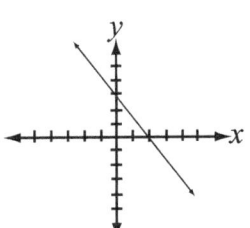

J.
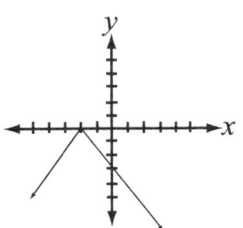

K.
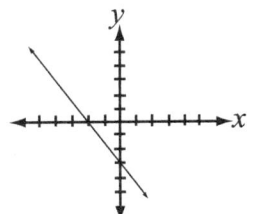

Absolute Value
Geometry Problem Set 9

11. On the number line below, which of the following is the best approximation of $|a + b|$?

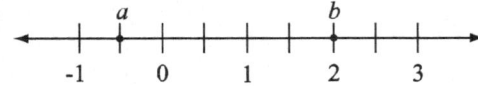

A. 0.5
B. 1.5
C. 2
D. 2.5
E. 3

Absolute Value
Geometry Problem Set 9

Answer Key

#	Answer	Frequency	Difficulty
1	B	popular	3
2	F	average	2
3	D	average	2
4	J	average	3
5	C	average	2
6	F	average	1
7	D	popular	1
8	H	popular	3
9	D	average	3
10	J	average	2
11	B	popular	2

Functions
Geometry Problem Set 10

1. Based on the portions of the graphs of the functions f and g shown below, what are all values of x between -5 and 5 for which $g(x) > f(x)$?

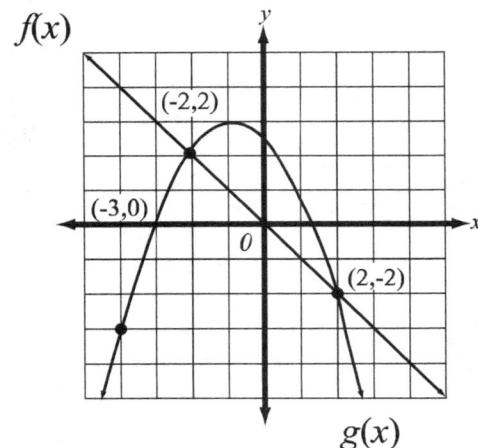

A. $-5 < x < 2$ only

B. $-2 < x < 2$ only

C. $-5 < x < -2$ and $2 < x < 5$

D. $-3 < x < 2$ only

E. $-5 < x < 5$ only

2. The functions f and g are linear functions, as shown on the graph below. What is the value of $f(2) + g(1)$?

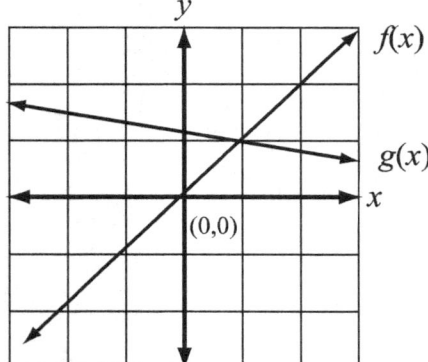

F. -1

G. 0

H. 1

J. 2

K. 3

Functions
Geometry Problem Set 10

3. Let the function $g(x)$ be defined as $g(x) = 3x + 6a$, where a is a constant. If $g(8) + g(12) = 42$, what is the value of a?

   A. $\dfrac{-3}{2}$

   B. $\dfrac{-2}{3}$

   C. $\dfrac{2}{3}$

   D. 1

   E. $\dfrac{3}{2}$

DO YOUR FIGURING HERE

4. The graph of $y = f(x)$ is shown below. If $0 < x < 10$, for how many values of x does $f(x) = 3$?

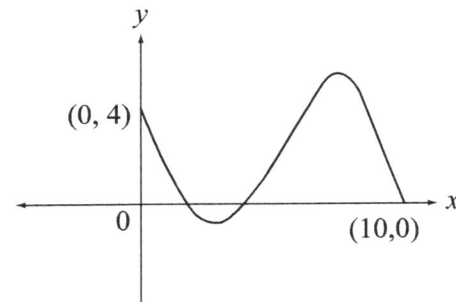

   F. None

   G. One

   H. Two

   J. Three

   K. More than three

Functions
Geometry Problem Set 10

5. Alan drove his car from his house to work one day. Along the way, he stopped for breakfast and then had to drive much faster than before to get to work on time. What of the following graphs could show the distance Alan traveled from his house as a function of time?

A.
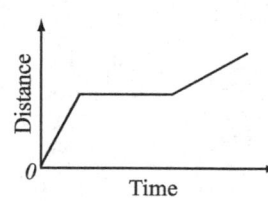

B.
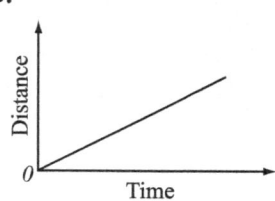

C.
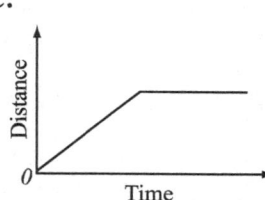

D.
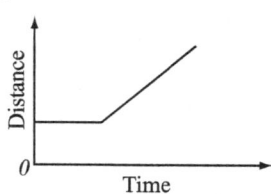

E.
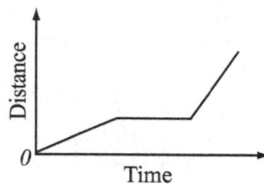

6. If $g(x) > 0$ for all real values of x, which of the following could be the equation of the function g?

   F. $g(x) = x + 1$

   G. $g(x) = x^2 + 1$

   H. $g(x) = x^2 - 1$

   J. $g(x) = x^3 + 1$

   K. $g(x) = x^3 - 1$

Functions
Geometry Problem Set 10

7. Which of the following is a graph of a function f such that $f(x) = 0$ for exactly two values of x between -5 and 5?

A.
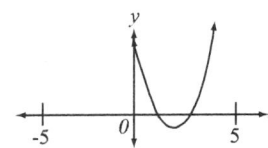

B.
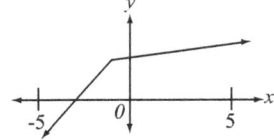

C.
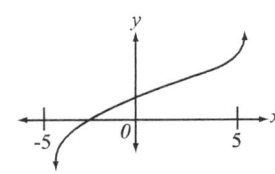

D.
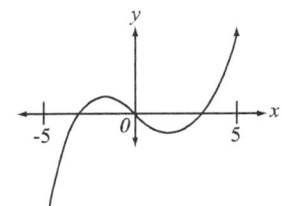

E.
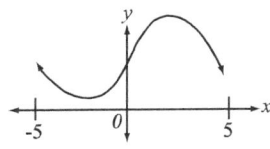

8. Let the functions f, g, and h be defined in the following ways: $f(x) = x^3$, $g(x) = x$, and $h(x) = f(x) - g(x)$. For $x > 1$, which of the following describes what happens to $h(x)$ as x increases?

   F. $h(x)$ stays the same

   G. $h(x)$ increases only

   H. $h(x)$ decreases only

   J. $h(x)$ increases at first, then decreases

   K. $h(x)$ decreases at first, then increases

Functions
Geometry Problem Set 10

9. The figure below shows the graph of $y = f(x)$, where f is a function. If $f(a) = f(3a)$, which of the following could be the value of a?

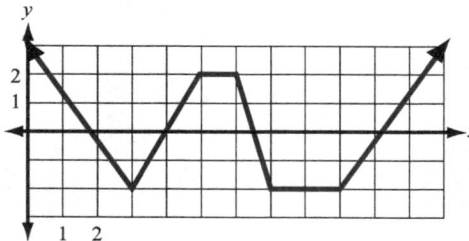

A. 1
B. 2
C. 3
D. 4
E. 5

DO YOUR FIGURING HERE

10. The graph below shows the function g, where $g(x) = k(x+3)(x-3)$ for some constant k. If $g(a - 1.5) = 0$ and $a > 0$, what is the value of a?

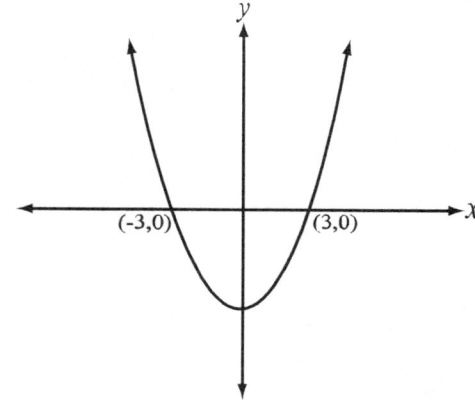

11. $$f(x) = 2x - 6$$
For the function defined above, what is one possible value of x for which $f(x) < x$?

Functions
Geometry Problem Set 10

12. In the xy-plane, the graph of a function f is a line. If $f(3) = 5$ and $f(11) = 13$, what is the value of $f(7)$?

F. 6.2

G. 8.6

H. 9

J. 9.8

K. 10

DO YOUR FIGURING HERE

Functions
Geometry Problem Set 10

Answer Key

#	Answer	Frequency	Difficulty
1	B	average	3
2	K	average	2
3	A	average	2
4	J	average	2
5	E	average	3
6	G	average	2
7	A	average	2
8	G	average	3
9	C	average	3
10	4.5	average	2
11	$x < 6$	average	3
12	H	average	4

Function Shifts
Geometry Problem Set 11

1. $f(x) = (x - 1)^2$. Which of the following graphs shows $g(x) = f(x) + 1$?

A.

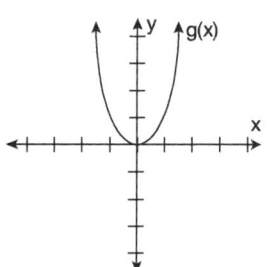

B.

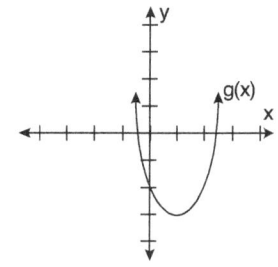

C.

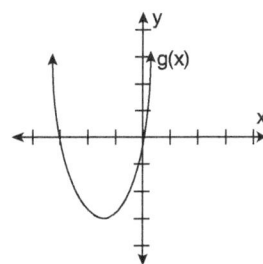

D.

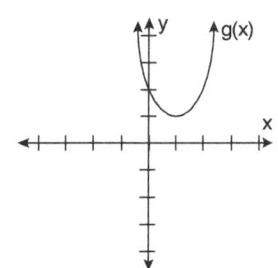

E.

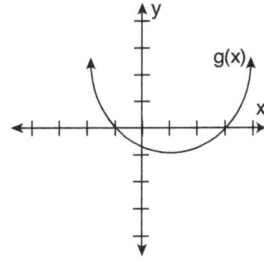

DO YOUR FIGURING HERE

Function Shifts
Geometry Problem Set 11

2. The graphs of the functions f and g are shown below over the interval from $x = -4$ to $x = 4$. Which of the following is one possible way to express f in terms of g?

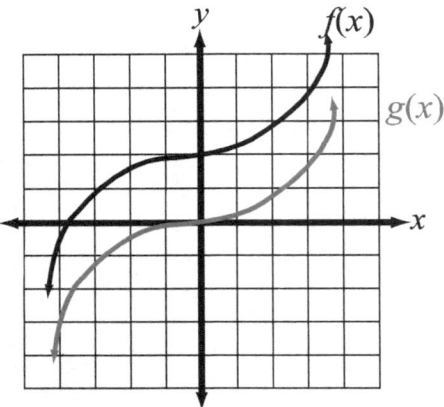

F. $f(x) = g(x+2)$

G. $f(x) = g(x+1) - 1$

H. $f(x) = g(x) + 1$

J. $f(x) = g(x) + 2$

K. $f(x) = g(x) - 2$

3. The functions $y = \sin x$ and $y = \sin(x+a) + b$, for constants a and b, are graphed in the standard xy-coordinate plane below. The functions have the same maximum value. Which of the following statements about the values of a and b could be true?

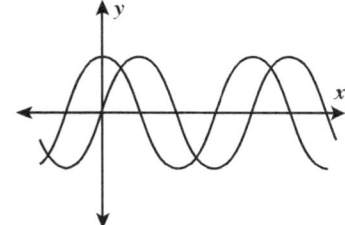

A. $a < 0$ and $b = 0$

B. $a < 0$ and $b > 0$

C. $a = 0$ and $b > 0$

D. $a > 0$ and $b < 0$

E. $a > 0$ and $b > 0$

Function Shifts
Geometry Problem Set 11

4. The figures below show the graphs of the functions f and g. Which of the following choices is a possible equation for $g(x)$?

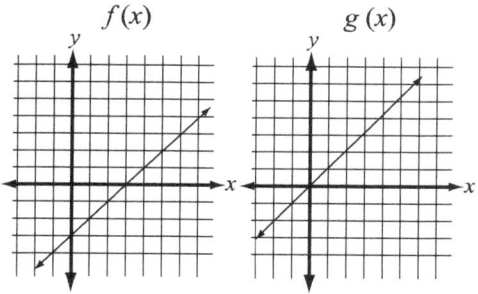

F. $f(x) + 2$

G. $f(x+1)$

H. $f(x) + 3$

J. $f(x+2)$

K. $f(x-2) - 1$

5. $f(x)$, shown below, shifts to form $g(x)$. If $g(3) = 2$, then what is one possible equation for $g(x)$?

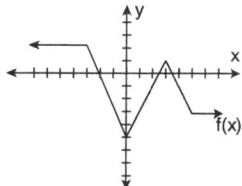

A. $f(x-5) + 2$

B. $f(x) + 2$

C. $f(x-4) - 1$

D. $f(x) - 1$

E. $f(x+2)$

Function Shifts
Geometry Problem Set 11

6. The figures below show the graphs of the functions $f(x)$ and $g(x)$. Which of the following is a possible equation for $g(x)$?

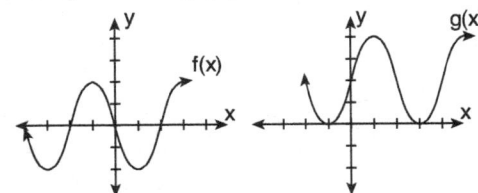

F. $f(x+1)+1$

G. $f\left(\dfrac{1}{2}x\right) - 2$

H. $f(x-2)+2$

J. $f(2x)+2$

K. $f(x)+4$

7. $f(x)$ is shown below. If $g(x) = f(x-3)+6$, then what is $g(5)$?

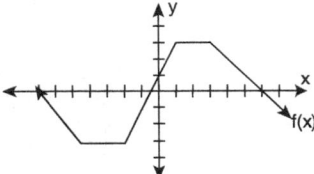

A. 5

B. 6

C. 8

D. 9

E. 11

8. $f(x) = 3x^2 + 2$. If $g(x) = f(x+2) - 3$, then what is the value of $g(6)$?

F. 110

G. 152

H. 178

J. 191

K. 194

Function Shifts
Geometry Problem Set 11

9. $f(x) = x^2$ and $g(x)$ are shown in the figures below. What is a possible way to express $g(x)$ in terms of $f(x)$?

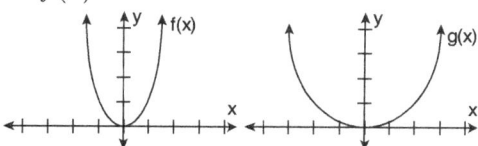

A. $f(x) + 10$

B. $f(2x)$

C. $f\left(\dfrac{1}{2}x\right)$

D. $2f(x)$

E. $(f(x))^2$

10. The figures below show the graphs of the functions f and g. The function f is defined by $f(x) = x^3 - 3x$. The function g is defined by $g(x) = f(x - h) + k$, where h and k are constants. What is the value of hk?

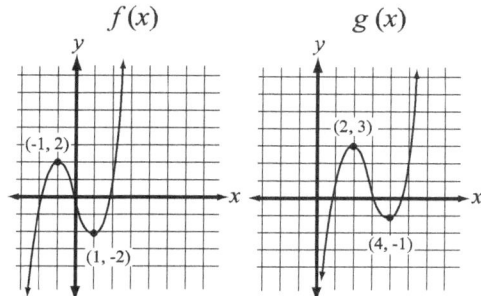

F. -6

G. -3

H. 1

J. 3

K. 6

Function Shifts
Geometry Problem Set 11

11. The figure below shows the graph of $y = g(x)$. If the function h is defined by $h(x) = g(3x) + 2$, what is the value of $h(2)$?

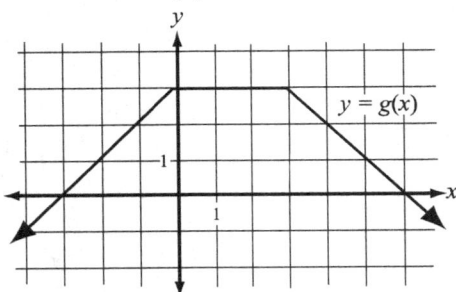

A. 2
B. 3
C. 5
D. 7
E. 11

DO YOUR FIGURING HERE

Function Shifts
Geometry Problem Set 11

Answer Key

#	Answer	Frequency	Difficulty
1	D	average	2
2	J	average	2
3	A	popular	1
4	H	average	2
5	A	average	2
6	H	average	2
7	D	average	2
8	J	average	3
9	C	average	4
10	J	average	4
11	A	average	4

Quadratics
Geometry Problem Set 12

1. Which of the following is the equation of the graph below?

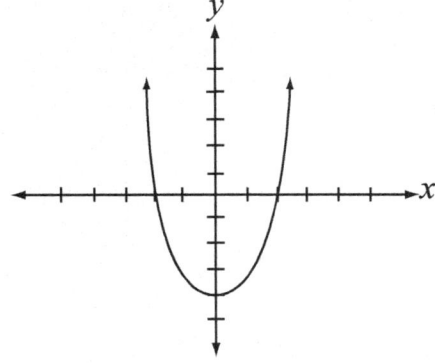

A. $y = (x - 4)^2$
B. $y = (x + 4)^2$
C. $y = x^2 + 4$
D. $y = x^2 - 4$
E. $y = -4x^2$

Quadratics
Geometry Problem Set 12

2. If the function f is defined by $f(x) = ax^2 + bx + c$, where a is positive and c is negative, which of the following could be the graph of f?

F.
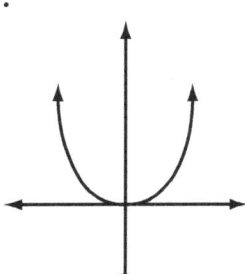

G.
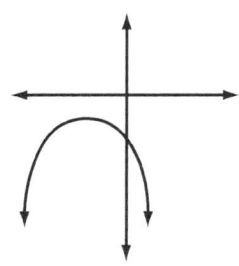

H.
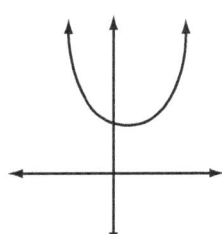

J.
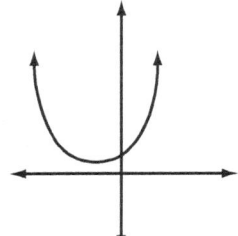

K.
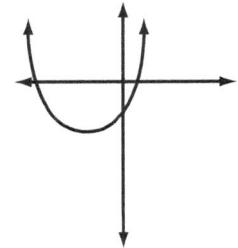

DO YOUR FIGURING HERE

Quadratics
Geometry Problem Set 12

3. The graph below is a parabola whose equation is $y = mx^2 - 3$, where m is a constant. If $y = -mx^2 - 3$ is graphed on the same axes, which of the following best describes the resulting graph as compared with the graph above?

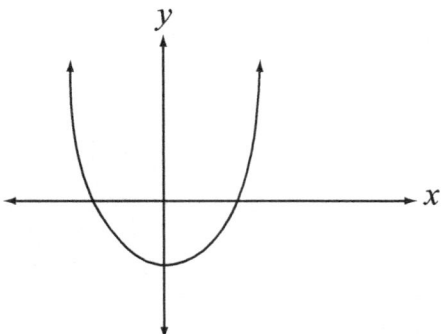

- A. It will be wider
- B. It will be narrower
- C. It will be moved down 3
- D. It will be moved to the right 3
- E. It will open downwards

4. If $x = 3$ is a solution of the equation $2x^2 + x + d = 0$, what is the value of d?
- F. -21
- G. -3
- H. 1
- J. 3
- K. 21

5. $$a^2 + 10a + b = (a + c)^2$$
In the equation above, b and c are constants. If the equation is true for all values of a, what is the value of b?
- A. -5
- B. 5
- C. 10
- D. 20
- E. 25

Quadratics
Geometry Problem Set 12

6. If $y = x^2 - 9$ and x is a real number, which of the follwing CANNOT be the value of y?

F. -10

G. -8

H. -4

J. 4

K. 10

7. If $x^2 + tx + 30 = (x + a)(x + 10)$ for all real values of x and if t and a are constants, what is the value of t?

A. 3

B. 5

C. 8

D. 13

E. 15

8. If b and c are constants and $x^2 + bx + 11$ is equivalent to $(x + 1)(x + c)$, what is the value of b?

F. 0

G. 10

H. 11

J. 12

K. It cannot be determined from the information given.

9.
$$x^2 + nx + 9 = (x + b)^2$$
In the equation above, n and b are positive constants. If the equation is true for all values of x, what is the value of n?

A. 2

B. 3

C. 6

D. 9

E. 12

Quadratics
Geometry Problem Set 12

10. The quadratic formula gives the 2 roots $x = \dfrac{-b \pm \sqrt{b^2 - 4ac}}{2a}$ for the equation $ax^2 + bx + c = 0$. What are the 2 roots for the equation $x^2 - x = 12$?

   F. $\dfrac{1 \pm \sqrt{-13}}{2}$

   G. $\dfrac{1 \pm \sqrt{13}}{2}$

   H. -4 and 3

   J. -3 and 4

   K. 2 and $\dfrac{3}{2}$

11. For a certain quadratic equation, $ax^2 + bx + c = 0$, the two solutions are $x = \dfrac{5}{3}$ and $x = -2$. Which of the following could be factors of $ax^2 + bx + c = 0$?

   A. $(3x - 5)$ and $(4x - 8)$

   B. $(3x - 5)$ and $(4x + 8)$

   C. $(3x + 5)$ and $(4x - 8)$

   D. $(3x - 5)$ and $(4x - 2)$

   E. $(3x - 5)$ and $(4x + 2)$

12. The figure below shows the graph of a quadratic function. What is the value of a?

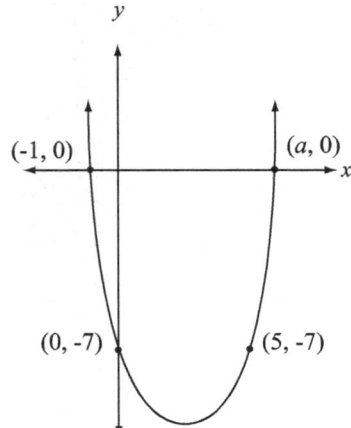

DO YOUR FIGURING HERE

Quadratics
Geometry Problem Set 12

13. The figure below shows the graph of a quadratic function in the xy-plane. Of all the points (x, y) on the graph, for what value of x is y the smallest?

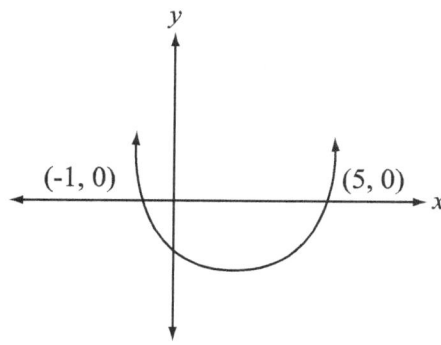

14. In the xy-coordinate plane, the graph of $x^2 = y + 3$ intersects with line m at $(a, 1)$ and $(b, 13)$. What is the greatest possible value of the slope of line m?

15. In the figure below, $ABCD$ is a square and points A, B, and P lie on the graph of $y = mx^2 + 3$, where m is constant. If the area of the square is 36, what is the value of m?

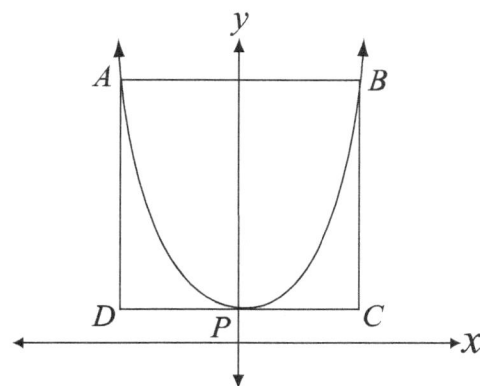

Quadratics
Geometry Problem Set 12

Answer Key

#	Answer	Frequency	Difficulty
1	D	average	3
2	K	average	3
3	E	average	2
4	F	average	3
5	E	rare	2
6	F	rare	2
7	D	rare	3
8	J	rare	2
9	C	rare	2
10	J	popular	2
11	B	popular	2
12	6	average	3
13	2	average	2
14	6	average	3
15	$\frac{2}{3}$	average	4

Circle Formula
Geometry Problem Set 13

1. Which of the following is the equation for a circle with the center (-2, -2) and which passes through the point (1, 2)?

   A. $(x - 2)^2 + (y - 2)^2 = 5$
   B. $(x - 2)^2 + (y - 2)^2 = 25$
   C. $(x + 2)^2 + (y + 2)^2 = 5$
   D. $(x + 2)^2 + (y + 2)^2 = 25$
   E. $(x + 2)^2 + (y + 2)^2 = 125$

2. In the xy-coordinate plane, a line intersects circle C at $(0, 2)$, at the center of circle C, and at $(6, 2)$. Which of the following is an equation for circle C?

   F. $(x - 2)^2 + (y - 3)^2 = 3$
   G. $(x - 3)^2 + (y - 2)^2 = 3$
   H. $(x - 3)^2 + (y - 2)^2 = 6$
   J. $(x - 2)^2 + (y - 3)^2 = 9$
   K. $(x - 3)^2 + (y - 2)^2 = 9$

3. In the xy-coordinate plane below, right triangle $\triangle ABC$ is inscribed in circle K, which is centered on the origin. $\overline{AB} = 12$, $\overline{AC} = 5$, and $\overline{BC} = 13$. Which of the following is an equation for circle K?

   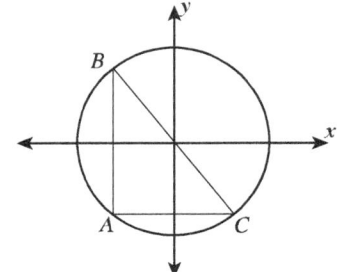

   A. $x^2 + y^2 = 2.5$
   B. $x^2 + y^2 = 6.5$
   C. $x^2 + y^2 = 12$
   D. $x^2 + y^2 = 42.25$
   E. $x^2 + y^2 = 169$

Circle Formula
Geometry Problem Set 13

4. A circle in the standard xy-coordinate plane is tangent to the x-axis at $x = 6$ and tangent to the y-axis at $y = 6$. Which of the following is an equation of the circle?

   F. $x^2 + y^2 = 6$

   G. $x^2 + y^2 = 36$

   H. $(x - 3)^2 + (y - 3)^2 = 9$

   J. $(x - 6)^2 + (y - 6)^2 = 36$

   K. $(x + 6)^2 + (y + 6)^2 = 36$

5. A circle has the equation $(x + 4)^2 + (y - 4)^2 = 4$. What is the center of this circle in the standard xy-coordinate plane?

   A. $(4, 4)$

   B. $(-4, 4)$

   C. $(4, -4)$

   D. $(-\sqrt{4}, \sqrt{4})$

   E. $(\sqrt{4}, -\sqrt{4})$

6. Which of the following has a center of $(1, 8)$ and a radius of 12?

   F. $x^2 + y^2 = 81$

   G. $(x + 1)^2 + (y + 8)^2 = 12$

   H. $(x - 1)^2 + (y - 8)^2 = 12$

   J. $(x + 1)^2 + (y + 8)^2 = 144$

   K. $(x - 1)^2 + (y - 8)^2 = 144$

7. Which of the following points could be on a circle described by the equation $(x - 4)^2 + (y + 2)^2 = 16$

   A. $(-12, 2)$

   B. $(-2, 4)$

   C. $(4, 2)$

   D. $(4, 18)$

   E. $(20, 2)$

Circle Formula
Geometry Problem Set 13

8. In the standard xy-coordinate plane below, a circle is inscribed in a square with vertices at $(0,0)$, $(8,0)$, $(8,8)$, and $(0,8)$. Which of the following is an equation of the circle shown?

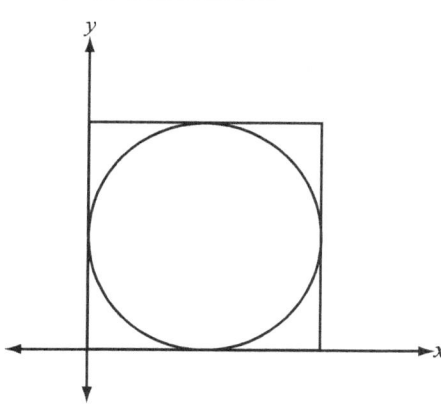

F. $x^2 + y^2 = 16$

G. $(x+8)^2 + (y+8)^2 = 64$

H. $(x-8)^2 + (y-8)^2 = 64$

J. $(x+4)^2 + (y+4)^2 = 16$

K. $(x-4)^2 + (y-4)^2 = 16$

9. Circle A is described by the equation $(x-3)^2 + (y-3)^2 = 9$. Circle B is tangent to circle A and has a radius that is $\frac{1}{3}$ that of circle A. Which of the following is the equation of circle B?

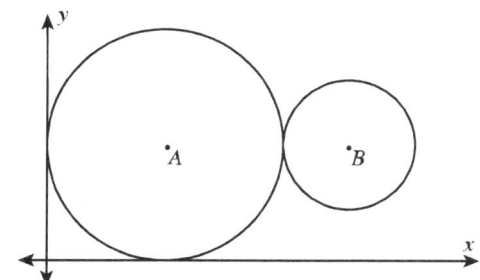

A. $\dfrac{(x-3)^2}{3} + \dfrac{(y-3)^2}{3} = 9$

B. $(x-3)^2 + (y-3)^2 = 1$

C. $(x-3)^2 + (y-3)^2 = 3$

D. $(x-7)^2 + (y-3)^2 = 1$

E. $(x-7)^2 + (y-3)^2 = 3$

Circle Formula
Geometry Problem Set 13

10. In the xy-coordinate plane, the points $A\,(1,4)$, $B\,(6,-1)$, and $C\,(1,-6)$ lie on a circle with center O. What are the coordinates of point O?

    F. $(0,0)$

    G. $(-1,1)$

    H. $(1,-1)$

    J. $(-4,-1)$

    K. $(1,0)$

DO YOUR FIGURING HERE

Circle Formula
Geometry Problem Set 13

Answer Key

#	Answer	Frequency	Difficulty
1	D	rare	2
2	K	rare	3
3	D	rare	2
4	J	rare	3
5	B	rare	2
6	K	rare	1
7	C	rare	2
8	K	rare	3
9	D	rare	4
10	H	rare	2

Intersections
Geometry Problem Set 14

1. If $y = 5$ intersects a quadratic function $f(x)$, at $x = 3$ and $x = 7$, then which of the following could be the equation for $f(x)$?

   A. $f(x) = x^2 + 5$

   B. $f(x) = (x - 5)^2 + 1$

   C. $f(x) = (x - 5)^2 + 5$

   D. $f(x) = (x + 1)^2 + 5$

   E. $f(x) = (x + 5)^2 + 1$

2. In the xy-coordinate plane, $(\sqrt{11}, a)$ is one of the points of intersection of the graphs of $y = 2x^2 - 18$ and $y = x^2 + b$, where b is a constant. What is the value of b?

   F. -7

   G. -4

   H. 0

   J. 4

   K. 7

3. The figure below shows the graphs of $y = x^2$ and $y = a - x^2$ for some constant a. If the length of $\overline{AB}$ is equal to 4, what is the value of a?

   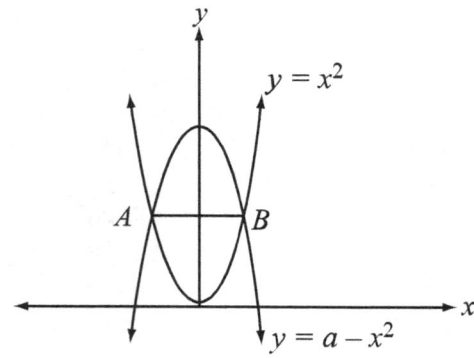

   A. 8

   B. 12

   C. 16

   D. 18

   E. 32

Intersections
Geometry Problem Set 14

4. In the xy-plane, line l passes through the origin and is perpendicular to the line $3x + y = k$, where k is a constant. If the two lines intersect at the point $(t, t+1)$, what is the value of t?

   F. $-\dfrac{3}{2}$

   G. $-\dfrac{2}{3}$

   H. $-\dfrac{1}{2}$

   J. $\dfrac{1}{2}$

   K. $\dfrac{3}{2}$

5. In the xy-coordinate plane, line l ($y = 3x - 2$) and line m ($y = -4x + 12$) intersect at point A. What are the coordinates of point A?

6. At what point(s) do $y = 3x^2$ and $y = 5x + 12$ intersect?

7. What is the equation (in slope-intercept form) of a line that intersects $y = x^2 - 5$ at $x = -3$ and $x = 2$?

Intersections
Geometry Problem Set 14

8. What are the (x, y) coordinates of the point of intersection between $3 = 2x - y$ and $-12 = x - 2y$?

9. In the xy-coordinate plane, the graph of $x = y^2 - 9$ intersects line l in the first quadrant at the points $(0, j)$ and $(7, k)$. What is the value of the slope of line l?

10. The xy-coordinate plane below shows a point of intersection, A, between a linear function and a quadratic function. What are the coordinates of point A?

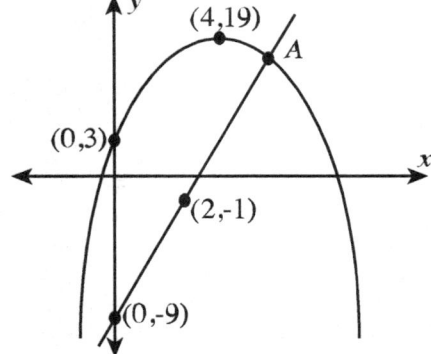

Intersections
Geometry Problem Set 14

Answer Key

#	Answer	Frequency	Difficulty
1	B	average	2
2	F	average	2
3	A	average	3
4	F	average	3
5	$(2, 4)$	average	1
6	$(3, 27)$ and $\left(-\frac{4}{3}, \frac{16}{3}\right)$	average	1
7	$y = -x + 1$	average	2
8	$(6, 9)$	average	2
9	$\frac{1}{7}$	average	3
10	$(6, 15)$	average	4

Logic
Geometry Problem Set 15

1. If it is raining or it is over 110° F, then the baseball game is cancelled. If the baseball game is not cancelled, which of the following is necessarily true?

    I. It is raining.
    II. It is not raining.
    III. It is not over 110° F.

    A. I only.
    B. II only.
    C. III only
    D. I and II only.
    E. II and III only.

2. At a bakery, all cupcakes are chocolate and gluten-free. All macarons are vanilla and lactose-free. If Katie purchases a baked good that is chocolate and gluten-free, which of the following is necessarily true?

    F. Katie purchased a macaron.
    G. Katie purchased a cupcake.
    H. Katie did not purchase a macaron.
    J. Katie did not purchase a cupcake.
    K. None of the above.

3. If it is sunny, then John will go to the beach or he will play soccer. If he plays soccer, he will see his friend Kathy. If he does not see his friend Kathy, which of the following must be true?

    I. It is not sunny.
    II. He does not play soccer.
    III. He goes sailing.

    A. I only.
    B. II only.
    C. III only.
    D. II and II only.
    E. I, II, and III.

DO YOUR FIGURING HERE

Logic
Geometry Problem Set 15

4. Gabi says, "If I walk slowly, I will be late." If this is true, which other statement must be true?

   F. If Gabi is late, then she walked slowly.

   G. If Gabi is not late, then she did not walk slowly.

   H. If Gabi walks quickly, then she will be early.

   J. If Gabi is not late, then she walked slowly.

   K. Gabi is late because she walked slowly.

5. All of the computers at Laney's school have a word processor.
   If the above statement is true, which of the following statements must also be true?

   A. If a computer does not have a word processor, then it is not at Laney's school.

   B. If a computer has a word processor, then it is at Laney's school.

   C. If a computer does not have a word processor, then it is at Laney's school.

   D. If a computer is not at Laney's school, then it doesn't have a word processor.

   E. Laney only uses a computer if it has a word processor.

6. A square is a rectangle.
   A rectangle is a quadrilateral.
   A rhombus is a quadrilateral.
   A square is a rhombus.
   Given that all of the statements above are true, which of the following must be true?

   F. All rhombuses are squares.

   G. All quadrilaterals are rectangles.

   H. All squares are rhombuses.

   J. All rhombuses are rectangles

   K. All rectangles are squares.

Logic
Geometry Problem Set 15

7. If it is Tuesday and there is no staff meeting, then Julia does not eat pizza.
   If Julia is eating pizza, which of the following statements must also be true?

   A. It is not Tuesday and there is a staff meeting.
   B. It is not Tuesday or there is a staff meeting.
   C. It is Tuesday.
   D. There is no staff meeting.
   E. There is a staff meeting.

8. If it is the weekend, Jamie will read a book or watch a movie.
   Which of the following is logically equivalent to the above statement?

   F. If Jamie watches a movie, then it is the weekend.
   G. If Jamie reads a book, then it is not the weekend.
   H. If Jamie reads a book and watches a movie, then it is the weekend.
   J. If Jamie does not read a book or watch a movie, then it is not the weekend.
   K. If Jamie does not read a book and he does not watch a movie, then it is not the weekend.

9. No values for x are square numbers.
   No values for y are prime numbers.
   All values for z are multiples of 2.
   Based on the above statements, which of the following can be a value for x, y, and z?

   A. 11
   B. 12
   C. 13
   D. 16
   E. 17

Logic
Geometry Problem Set 15

10. Which of the following statements is a logical conclusion from the 3 true statements given below?

    All leeps are grools
    All parps are grools
    All krins are leeps

    F. No parps are grools
    G. No parps are krins
    H. All krins are grools
    J. All parps are leeps
    K. All leeps are parps

Logic
Geometry Problem Set 15

Answer Key

#	Answer	Frequency	Difficulty
1	E	rare	1
2	H	rare	1
3	B	rare	1
4	G	rare	1
5	A	rare	1
6	H	rare	1
7	B	rare	1
8	K	rare	1
9	B	rare	1
10	H	rare	3

Coordinate Geometry Mixed Problem Set
Geometry Problem Set 16

1. In the xy-coordinate plane, the distance from point $(x, 3)$ to point $(3, 7)$ is 5. What is a possible value of x?

   A. -6
   B. -1
   C. 0
   D. 2
   E. 4

2. In the xy-coordinate plane, point $P(0, -2)$ and point $R(8, 6)$ are opposite vertices of square $PQRS$. If point A is the midpoint of line $\overline{PS}$, then what is the area of $\triangle PAQ$?

   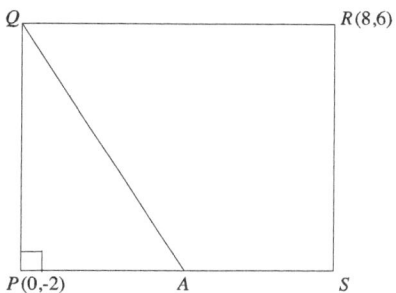

   F. 16
   G. 21
   H. 28
   J. 32
   K. 64

3. In the xy-coordinate plane, point A lies on the line $2x + 3(y - 8) = 0$. If the y-coordinate is 7, then what is the x-coordinate?

   A. $-\dfrac{2}{3}$
   B. 0
   C. $\dfrac{1}{2}$
   D. $\dfrac{3}{2}$
   E. 6

Coordinate Geometry Mixed Problem Set
Geometry Problem Set 16

4. In the xy-coordinate plane, line m has the equation $6 = -24x + 6y$. Line n is parallel to line m and passes through the point $(3, 2)$. What is the equation of line n?

   F. $y = -24x + 74$
   G. $y = 4x - 5$
   H. $y = 4x - 10$
   J. $y = 6x$
   K. $y = 8x - 5$

5. In the xy-coordinate plane, line $y = mx + 142$ is perpendicular to line $8y - 14x = 16$. What is the value of m?

   A. -7
   B. $-\dfrac{4}{7}$
   C. $\dfrac{1}{16}$
   D. $\dfrac{7}{4}$
   E. 4

6. In the xy-coordinate plane, $\triangle ABC$ has points $A(-2,-4)$, $B(-2,6)$, and $C(2,6)$. What is the area of $\triangle ABC$?

   F. 4
   G. 8
   H. 12
   J. 20
   K. 40

7. In the xy-coordinate plane, segment $\overline{AB}$ has a midpoint M. If A has the coordinates $(1, 8)$ and B has the coordinates $(1, 72)$, then what is the length of segment $\overline{AM}$?

   A. 4
   B. 32
   C. 40
   D. 64
   E. 80

DO YOUR FIGURING HERE

Coordinate Geometry Mixed Problem Set
Geometry Problem Set 16

8. In the xy-coordinate plane, points $A, B, C,$ and D lie on a line in that order. B is the midpoint of line $\overline{AD}$. If the length of $\overline{AD}$ is 12 and the length of line $\overline{BC}$ is 2, then what is the length of line $\overline{CD}$?

   F. 2
   G. 4
   H. 5
   J. 6
   K. 10

9. In the xy-coordinate plane, points $A\,(4,-2)$ and $B\,(0,-6)$ are two adjacent corners of the square $ABCD$. What is the area of square $ABCD$?

   A. 4
   B. 5
   C. 16
   D. 20
   E. 25

10. When graphed in the xy-coordinate plane, which of the following DOES NOT represent a line?

    F. $x = 0$
    G. $y = 4x + 18$
    H. $x^2 + y = 1$
    J. $\dfrac{1}{4}y = 4x$
    K. $x = y + 12$

11. What is the slope of line m if it is perpendicular to the line that passes through point $A\,(-1, 0)$ and point $B\,(4, -7)$?

    A. -7
    B. $-\dfrac{7}{3}$
    C. $-\dfrac{7}{5}$
    D. $\dfrac{5}{7}$
    E. 5

Coordinate Geometry Mixed Problem Set
Geometry Problem Set 16

12. Which of the following are equations for lines with a slope of $\frac{1}{4}$?

    I. $x = 4y + 8$
    II. $y = x^2 + x + 1$
    III. $4y = x - 13$

    F. I only

    G. II only

    H. III only

    J. I and III only

    K. I, II, and III

13. Line $\overline{AB}$ is 13 units long. Point A has the coordinates $(-2, 0)$ and point B has the coordinates $(3, y)$. What is a possible value of y?

    A. -12

    B. -6

    C. -2

    D. 4

    E. 5

14. A system of equations is shown below:
    $$6y = \frac{1}{4}x + 2$$
    $$6y = 4x + 2$$
    Which statement best describes the two lines?

    F. Two distinct, intersecting lines

    G. Two parallel lines

    H. Two perpendicular lines, with negative slopes

    J. Two perpendicular lines, with positive slopes

    K. A single line with a positive slope

Coordinate Geometry Mixed Problem Set
Geometry Problem Set 16

15. Point A has the y-coordinate 31 and lies on the parabola $y = 4x^2 - 2x + 1$. What is a possible x-coordinate of point A?

    A. -3
    B. 3
    C. 20
    D. 24
    E. 26

16. A line consists of points $ABCD$, in that order. Line $\overline{AD}$ is 16 units. B is the midpoint of segment $\overline{AD}$ and C is the midpoint of segment $\overline{BD}$. What is the length of $\overline{AC}$?

    F. 2
    G. 4
    H. 8
    J. 12
    K. 16

17. What is the range of the function defined as $f(x) = x^2 - 3$?

    A. All real numbers
    B. All integers
    C. $f(x) \geq 0$
    D. $g(x) \geq -3$
    E. $g(x) \geq 3$

18. Line A passes through the origin and point $B(4, 5)$. What is the slope of line A?

    F. 0
    G. $\dfrac{1}{5}$
    H. $\dfrac{4}{5}$
    J. $\dfrac{5}{4}$
    K. 4

Coordinate Geometry Mixed Problem Set
Geometry Problem Set 16

19. Which of the following shares an x-intercept with $4y + 2x = 3$?

A. $4y + x = 3$

B. $x - 3 = 2y$

C. $y = \frac{1}{2}x + \frac{3}{2}$

D. $x = y + \frac{3}{2}$

E. $2x = 4$

20. The function $y = x^2 + 2x - 3$ intercepts the x-axis at what point(s)?

F. $(3, 0)$

G. $(0, 1)$

H. $(-3, 0)$ and $(1, 0)$

J. $(0, -3)$ and $(0, 1)$

K. $(-5, 0)$ and $(5, 0)$

21. Point A is the point with the smallest y-coordinate on the semicircle below. If the coordinates of A are $(4.5, -6)$, what is the x-coordinate of point B?

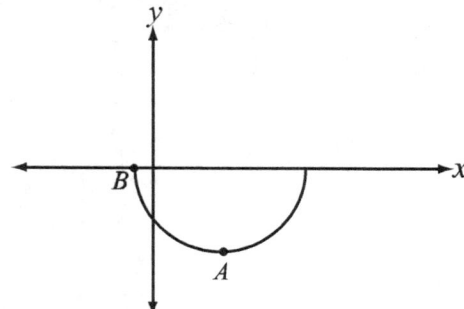

A. -2.5

B. -2

C. -1.5

D. -1

E. -.5

Coordinate Geometry Mixed Problem Set
Geometry Problem Set 16

22. In the xy-coordinate plane below, the circle has center $(3, 3)$. Which of the following lines will divide the circle into two semicircles?

    I. The line with equation $y = x$
    II. The line with equation $x = 3$
    III. The line with equation $y = 3$

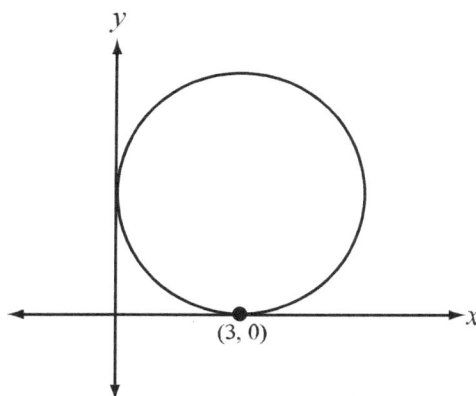

F. I only

G. II only

H. I and II only

J. II and III only

K. I, II, and III

Coordinate Geometry Mixed Problem Set
Geometry Problem Set 16

23. The equation of the line below is $y = \dfrac{3}{2}x + 3$. Which of the following is the graph of $y = -\left|\dfrac{3}{2}x + 3\right|$?

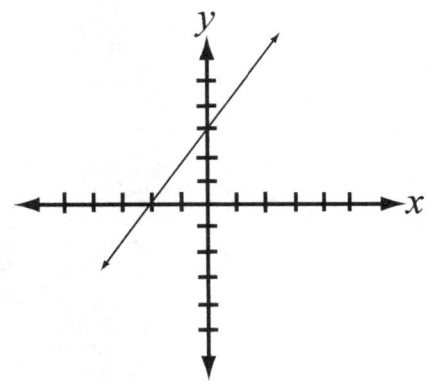

A.
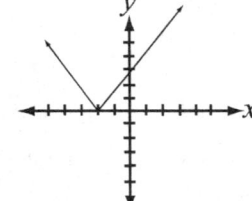

B.
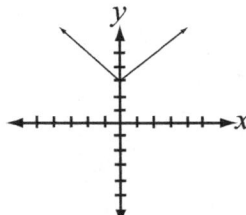

C.
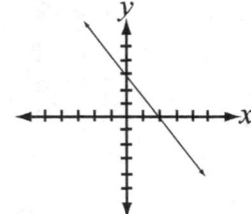

D.
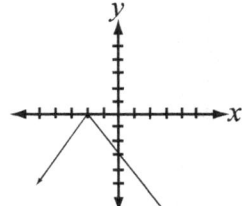

E.
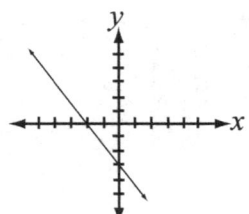

Coordinate Geometry Mixed Problem Set
Geometry Problem Set 16

24. Which of the points in the xy-plane shown below satisfy the equation $x + |y| = 2$?

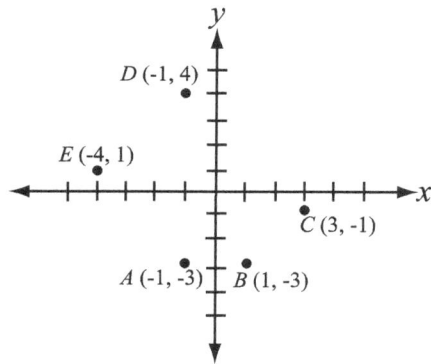

   F. A
   G. B
   H. C
   J. D
   K. E

25. Segment $\overline{AB}$ has midpoint M. If the length of $\overline{AM}$ is $4x$, what is the length of $\overline{AB}$ in terms of x?

   A. x
   B. $2x$
   C. $4x$
   D. $8x$
   E. $16x$

26. On the number line below, the tick marks are equally spaced. What is the value of $b - a$?

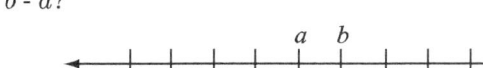

   F. 0.2
   G. 0.4
   H. 0.6
   J. 0.8
   K. 1

DO YOUR FIGURING HERE

Coordinate Geometry Mixed Problem Set
Geometry Problem Set 16

27. On the number line below, which of the following is the best approximation of $|a + b|$?

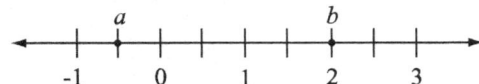

- A. 0.5
- B. 1.5
- C. 2
- D. 2.5
- E. 3

28. If x is the coordinate of the indicated point on the number line below, which of the lettered points has the coordinate $\frac{-1}{2}x$?

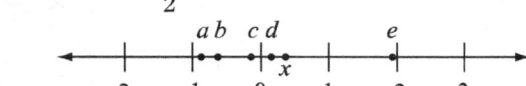

- F. A
- G. B
- H. C
- J. D
- K. E

29. The letters $a, b, c,$ and d represent numbers as shown on the number line below. Which of the following expressions has the highest value?

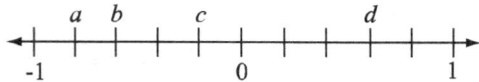

- A. $d + c$
- B. $d - b$
- C. $a - d$
- D. $a - b$
- E. $c - d$

Coordinate Geometry Mixed Problem Set
Geometry Problem Set 16

30. Points A, B, C, and D lie on a line in that order. C is the midpoint of $\overline{AD}$. If the length of $\overline{AD}$ is 18 and the length of $\overline{AB}$ is 7, what is the length of $\overline{BC}$?

F. 1
G. 2
H. 3
J. 4
K. 5

31. In the triangle below, $\overline{AC}$ has a slope of $\frac{-9}{22}$. If $\overline{AB}$ is 4.5, what is the length of $\overline{BC}$?

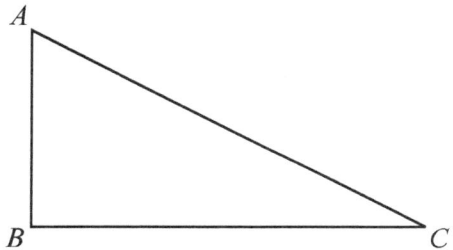

A. 11
B. 20.25
C. 22
D. 40.5
E. 44

Coordinate Geometry Mixed Problem Set
Geometry Problem Set 16

32. In the xy-coordinate plane below, the point $(a, 19)$ (not shown) lies on the line m. What is the value of a?

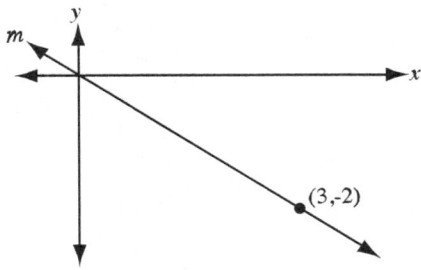

F. -28.5
G. -11
H. 11
J. 22
K. 28.5

33. In the xy-coordinate plane below, which of the following line segments has a slope of -2?

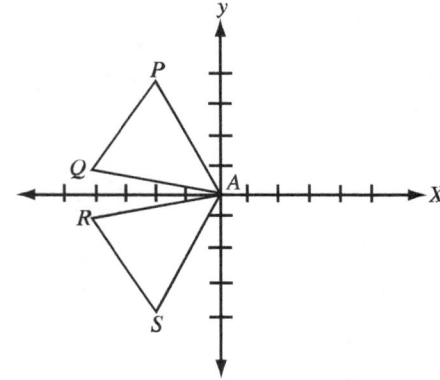

A. $\overline{AQ}$
B. $\overline{AP}$
C. $\overline{AR}$
D. $\overline{AS}$
E. $\overline{RS}$

Coordinate Geometry Mixed Problem Set
Geometry Problem Set 16

34. Tick marks are equally spaced on the number line below. Which of the lettered ticks has a coordinate equal to $\frac{(-2)^2}{5}$?

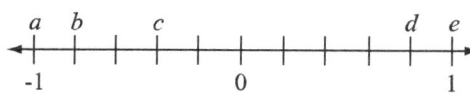

- **F.** a
- **G.** b
- **H.** c
- **J.** d
- **K.** e

35. In the figure below, what is the median (middle-most value) of the slopes of $\overline{AP}, \overline{AQ}, \overline{AR}, \overline{AS}$, and $\overline{AT}$?

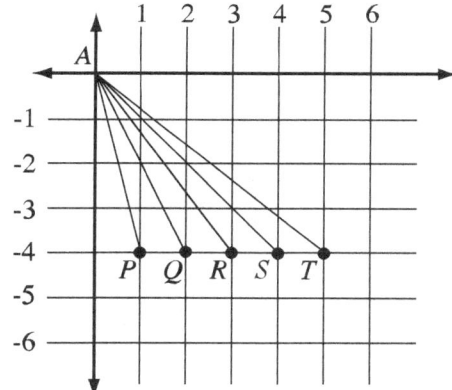

- **A.** $\frac{-4}{3}$
- **B.** -1
- **C.** $\frac{-3}{4}$
- **D.** $\frac{3}{4}$
- **E.** $\frac{4}{3}$

Coordinate Geometry Mixed Problem Set
Geometry Problem Set 16

36. What is the point on a number line that is exactly halfway between -8 and 47?

    F. 19.5
    G. 24
    H. 26.5
    J. 27.5
    K. 32.5

37. The graph below shows a parabola whose line of symmetry has the equation $x = 2$. If the x-intercepts of the parabola are located at $(n, 0)$ and $(5, 0)$, what is the value of n?

    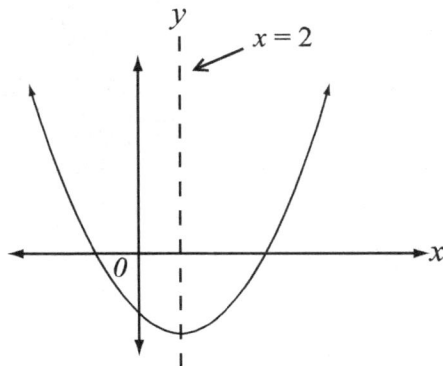

    A. -5
    B. -3
    C. -1
    D. 0
    E. 1

DO YOUR FIGURING HERE

Coordinate Geometry Mixed Problem Set
Geometry Problem Set 16

38. Jorge plans a day-long road trip, including breaks at rest stops. The graph below shows the relationship between time and total distance traveled during the road trip. What was Jorge's average speed, in miles per hour, for the parts of the day when he was traveling from one rest stop to another?

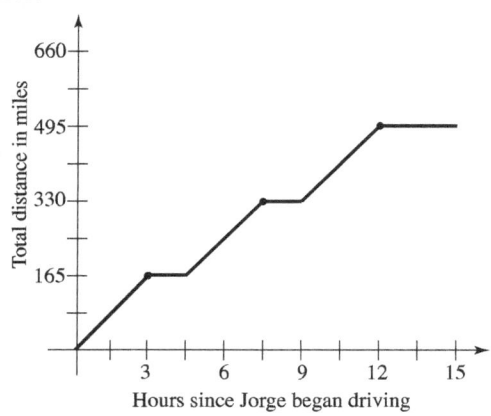

F. 33
G. 41
H. 45
J. 55
K. 165

Coordinate Geometry Mixed Problem Set
Geometry Problem Set 16

39. The *domain* of a function f is all values of x for which $f(x)$ is defined. Which of the following sets is the domain for the function graphed below?

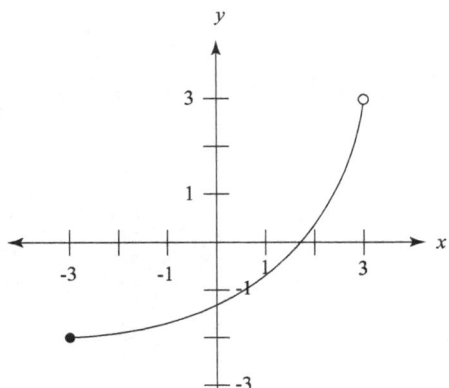

- **A.** $\{x : \text{-}3 \leq x \leq 3\}$
- **B.** $\{x : \text{-}3 \leq x < 3\}$
- **C.** $\{x : \text{-}2 \leq x \leq 2\}$
- **D.** $\{\text{-}2, \text{-}1, 0, 1, 2\}$
- **E.** $\{\text{-}3, \text{-}2, \text{-}1, 0, 1, 2, 3\}$

40. Kendra's house is 3 miles north and 16 miles west of Toby's house. Felix's house is 2 miles south and 12 miles east of Toby's house. Approximately how many miles apart are Kendra and Felix's houses?

- **F.** 8.1
- **G.** 27.5
- **H.** 28.0
- **J.** 28.4
- **K.** 33.0

Coordinate Geometry Mixed Problem Set
Geometry Problem Set 16

41. Line l ($4y = 2x + 3$) is reflected over the x-axis to form line m. What is the equation for line m?

A. $y = -\frac{1}{2}x - \frac{3}{4}$

B. $y = -2x - \frac{3}{4}$

C. $y = \frac{1}{2}x + \frac{3}{4}$

D. $y = 3x + \frac{1}{2}$

E. $y = x + 6$

42. In the figure below, line m passes through the origin. What is the value of $\frac{a}{b}$?

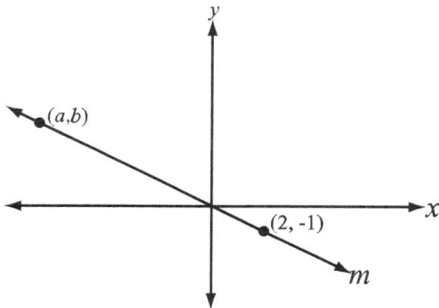

F. -2

G. $\frac{-1}{2}$

H. 0

J. $\frac{1}{2}$

K. 2

Coordinate Geometry Mixed Problem Set
Geometry Problem Set 16

43. What is the perimeter of the square in the figure below?

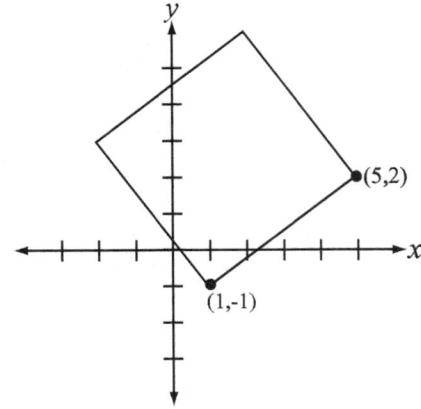

- A. 20
- B. 22
- C. 24
- D. 26
- E. 28

44. In the xy-coordinate plane, point A is $(-3, 2)$, point B is $(2, 2)$ and point C is $(-3, -4)$. What is the area of $\triangle ABC$?

- F. 5
- G. 10
- H. 12
- J. 15
- K. 30

Coordinate Geometry Mixed Problem Set
Geometry Problem Set 16

45. What is one possible equation for the graph shown below?

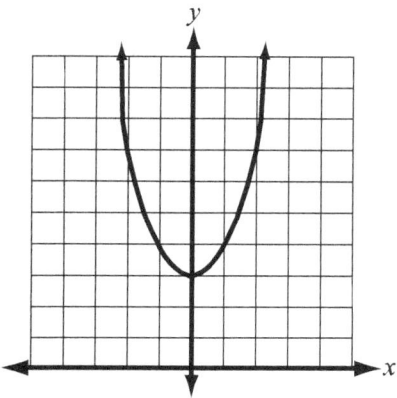

- **A.** $y = x^2 - 3$
- **B.** $y = 3x^2$
- **C.** $y = x^2 + 3$
- **D.** $y = x + 3$
- **E.** $y = 3x^2 + 3$

46. In the xy-coordinate plane below, the quadrants are labeled I through IV. Line m (not shown) does not contain points in either quadrant I or quadrant III. Which of the following could be the equation of line m?

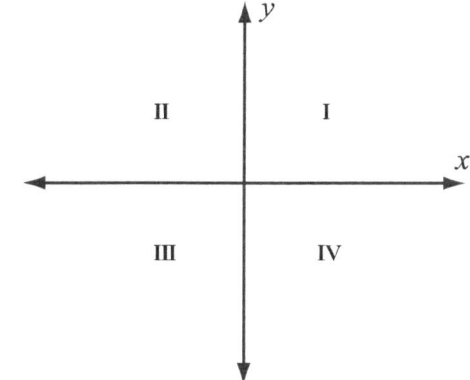

- **F.** $y = 1$
- **G.** $x = 1$
- **H.** $y = -3x$
- **J.** $y = -3x - 5$
- **K.** $y = 3x - 5$

DO YOUR FIGURING HERE

Coordinate Geometry Mixed Problem Set
Geometry Problem Set 16

47. Which of the following could be the equation of the graph of function f, shown below?

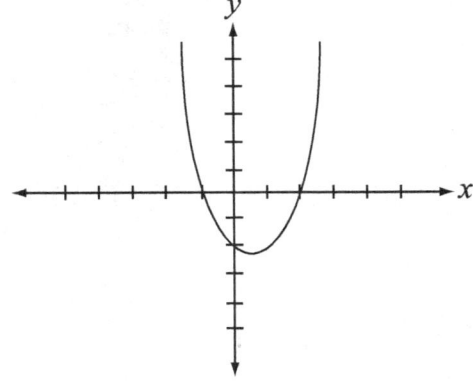

- **A.** $f(x) = x^2 + x + 3$
- **B.** $f(x) = x^2 + x - 3$
- **C.** $f(x) = x^2 - x + 3$
- **D.** $f(x) = x^2 - x - 2$
- **E.** $f(x) = x^2 - 3x + 1$

48. In the figure below, line m has a slope of $-\dfrac{1}{2}$, and the rectangle under the line has a height of 2 and a width of 6. What is the x-intercept of m?

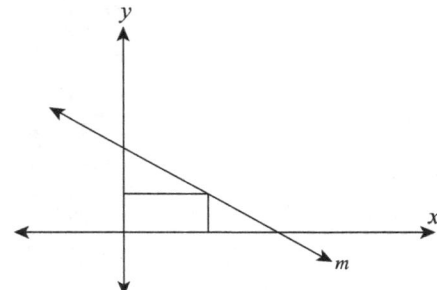

- **F.** 8
- **G.** 9
- **H.** 10
- **J.** 11
- **K.** 12

Coordinate Geometry Mixed Problem Set
Geometry Problem Set 16

49. A parallelogram is graphed in the xy-coordinate plane, with verticies $A\,(-2,3)$, $B\,(4,8)$, and $C\,(6,0)$. Which of the following are possible coordinates for the fourth vertex, point D?

  A. $(-2,0)$

  B. $(0,-5)$

  C. $(1,-3)$

  D. $(2,-6)$

  E. Cannot be determined

DO YOUR FIGURING HERE

50. The function $f(x)$ is shown below. Which of the following would shift $f(x)$ three units to the right and two units down?

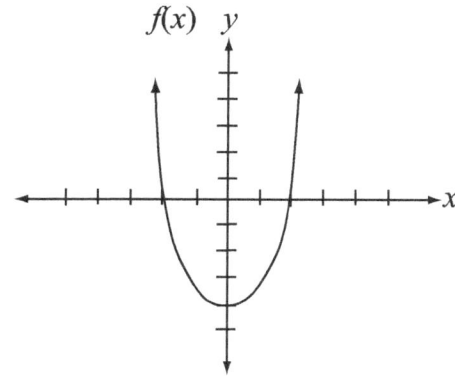

  F. $f(x-3)+2$

  G. $f(x-3)-2$

  H. $f(x+3)+2$

  J. $f(x+3)-3$

  K. $f(x+3)+3$

Coordinate Geometry Mixed Problem Set
Geometry Problem Set 16

51. The figure below shows the graph of the function $a(x)$. If $b(x) = 0.5(a(x))$ for all values of x, which of the following is a true statement describing the graph of $b(x)$ in comparison with the graph of $a(x)$?

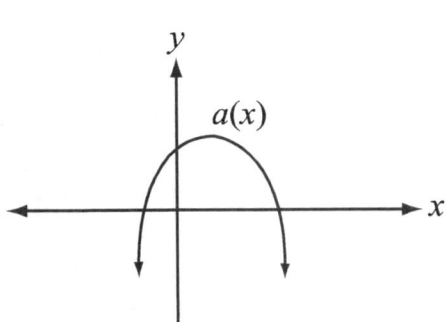

A. It has the same shape as the graph of a but opens upward

B. It is wider than the graph of a and opens downward

C. It is wider than the graph of a and opens upward

D. It is narrower than the graph of a and opens downward

E. It is narrower than the graph of a and opens upward

Coordinate Geometry Mixed Problem Set
Geometry Problem Set 16

52. In the figure below, $ABCD$ is a rectangle. Point D lies at $(-2, 0)$ and Point C lies at $(2, 0)$. Points A and B lie on the graph of $y = mx^2$, where m is a constant. If the perimeter of ABCD is 10, what is the value of m?

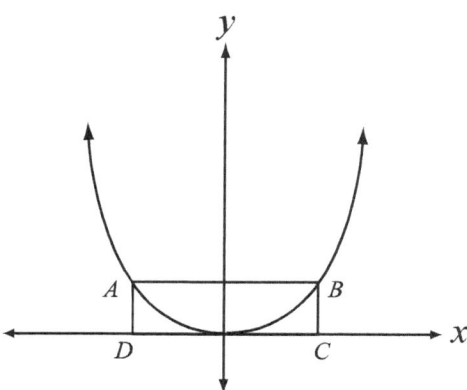

F. 0.25
G. 0.5
H. 0.75
J. 1
K. 2

Coordinate Geometry Mixed Problem Set
Geometry Problem Set 16

53. In the figure below, the graph of $y = 2x^2 + x - 10$ intersects the y-axis at point A and the x-axis at point C. What is the area of rectangle $ABCD$?

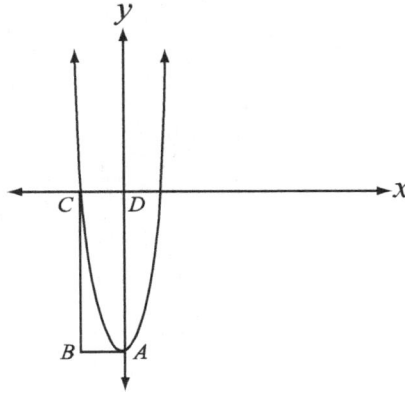

- **A.** 10
- **B.** 15
- **C.** 20
- **D.** 25
- **E.** 30

54. Points P, Q, R and S lie on a line in that order. $\overline{PR} = \dfrac{3}{8}\overline{PS}$ and $\overline{QS} = \dfrac{5}{6}\overline{PS}$. If $\overline{PS} = 24$, what is the length of $\overline{QR}$?

Coordinate Geometry Mixed Problem Set
Geometry Problem Set 16

Answer Key

#	Answer	Frequency
1	C	popular
2	F	popular
3	D	average
4	H	popular
5	B	popular
6	J	popular
7	B	popular
8	G	popular
9	C	popular
10	H	popular
11	D	popular
12	J	popular
13	A	popular
14	F	popular
15	B	popular
16	J	popular
17	D	popular
18	J	popular
19	D	average
20	H	average
21	C	average
22	K	average
23	D	average
24	F	average
25	D	popular
26	F	popular
27	B	popular
28	H	popular
29	B	popular
30	G	popular
31	A	popular
32	F	average
33	B	popular
34	J	popular
35	A	popular
36	F	popular
37	C	popular
38	J	rare
39	B	average
40	J	popular
41	A	average
42	F	popular
43	A	popular
44	J	popular
45	C	average

#	Answer	Frequency
46	H	popular
47	D	popular
48	H	popular
49	B	popular
50	G	average
51	B	average
52	F	popular
53	D	average
54	5	popular

Parallel Lines and Transversals
Geometry Problem Set 17

1. In the figure below, lines r, s, and t are parallel. If $a = 140$, what is the value of $b + c$?

   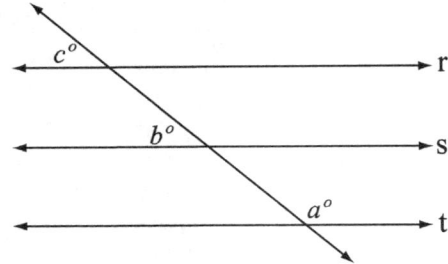

   DO YOUR FIGURING HERE

   A. 40
   B. 60
   C. 80
   D. 90
   E. 100

2. The figure below shows five lines. If $m \parallel n$, $x \parallel y$ and x is perpendicular to n which of the following is NOT equal to 90°?

   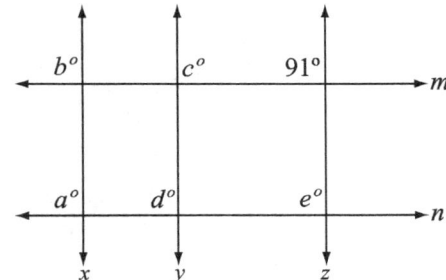

   F. a
   G. b
   H. c
   J. d
   K. e

Parallel Lines and Transversals
Geometry Problem Set 17

3. In the figure below, lines k and l are NOT parallel. If $a = 62$, which of the following CANNOT be the value of b?

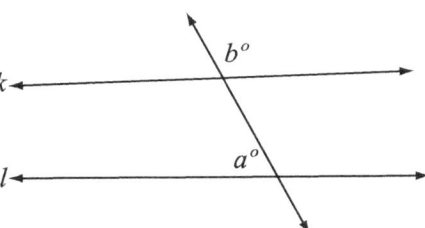

A. 117
B. 118
C. 119
D. 120
E. 121

4. In the figure below $\overline{AB}$ and $\overline{CD}$ are parallel. $\overline{JK}$ intersects $\overline{AB}$ at M and $\overline{CD}$ at N. Which of the following statements must be true?

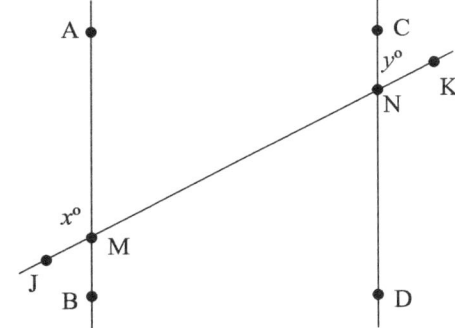

F. $x = y$
G. $x + y = 90$
H. $x - y = 90$
J. $x + y = 180$
K. $x - y = 180$

Parallel Lines and Transversals
Geometry Problem Set 17

5. In the figure below, lines i and j are parallel. Line k and line l are perpendicular. What is the value of x?

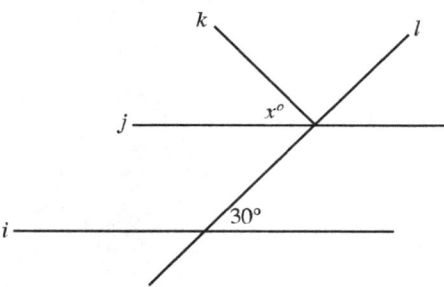

- A. 30
- B. 45
- C. 60
- D. 85
- E. 150

6. In the figure below, lines j and k are parallel to each other and line l is perpendicular to line k. Which of the following has the largest value?

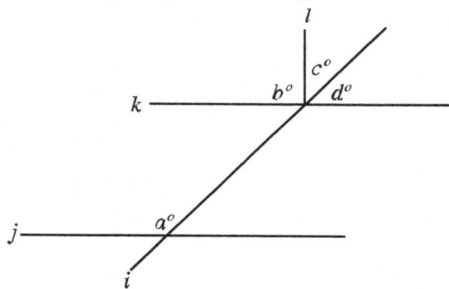

- F. a
- G. $a + d$
- H. $b + c$
- J. $b + d$
- K. $c + d$

Parallel Lines and Transversals
Geometry Problem Set 17

7. In the figure below, lines i and j are parallel. Which of the following are equivalent?

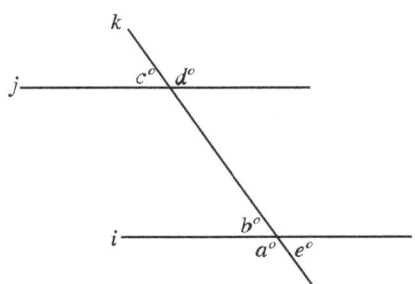

A. a, b, and e

B. a and c

C. b, c, and e

D. d and e

E. b, d, and e

8. In the figure below, lines i and j are parallel. What is the value of a?

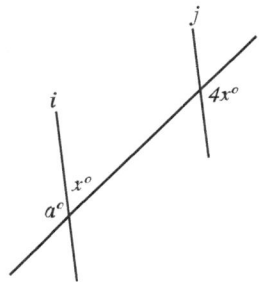

F. 36

G. 72

H. 88

J. 144

K. 120

Parallel Lines and Transversals
Geometry Problem Set 17

9. In the figure below, lines i and j are parallel. What is the value of a?

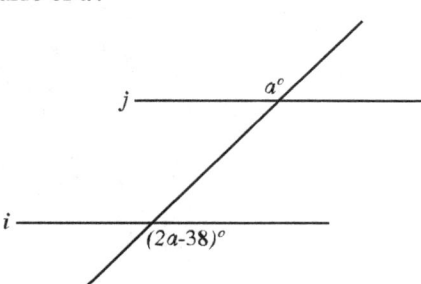

A. 32
B. 38
C. 68
D. 73
E. 109

10. In the figure below, lines i and j are parallel. What is the value of a?

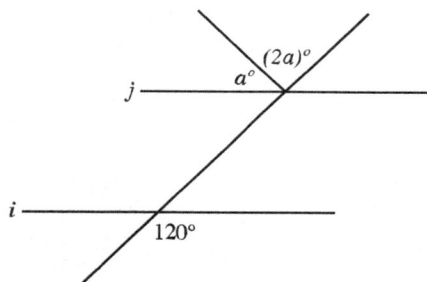

F. 20
G. 35
H. 40
J. 60
K. 80

Parallel Lines and Transversals
Geometry Problem Set 17

11. In the figure below, lines g and h are parallel.
    Lines i and j intersect g at the same point.
    $$\text{Angle } a = 4x + 12$$
    $$\text{Angle } b = 3x - 18$$
    $$\text{Angle } c = 5x - 18$$
    What is the value of x?

    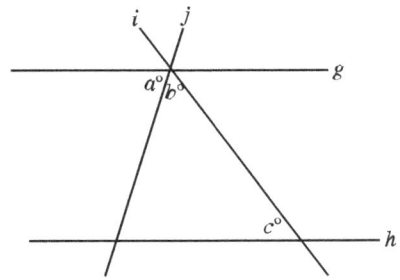

    A. 10
    B. 13
    C. 16
    D. 17
    E. 18

DO YOUR FIGURING HERE

Parallel Lines and Transversals
Geometry Problem Set 17

Answer Key

#	Answer	Frequency	Difficulty
1	C	popular	1
2	K	popular	1
3	B	popular	2
4	J	popular	1
5	C	average	1
6	G	popular	2
7	C	popular	1
8	J	popular	2
9	B	popular	1
10	H	popular	1
11	D	popular	3

Lines and Angles
Geometry Problem Set 18

1. For the two intersecting lines below, which of the following must be true?

   I. $b = 2a$
   II. $a > c$
   III. $c + 110 = a + b$

   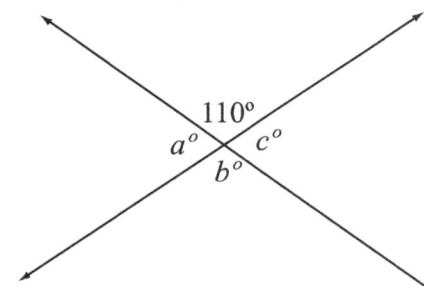

   A. II only
   B. III only
   C. II and III only
   D. I and III only
   E. I, II, and III

2. In the figure below, three lines intersect as shown. What is the value of a?

   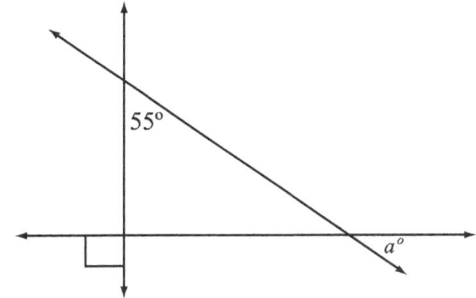

   F. 30
   G. 35
   H. 40
   J. 45
   K. 50

Lines and Angles
Geometry Problem Set 18

3. In the figure below, r is perpendicular to u. What is the value of a?

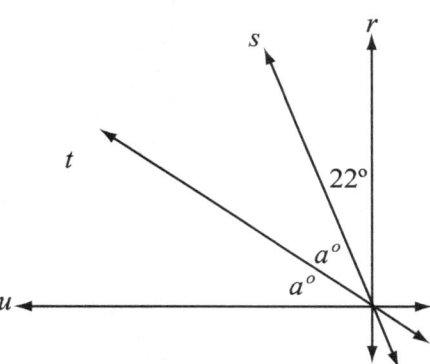

A. 22
B. 24
C. 34
D. 40
E. 45

4. In the figure below, point O lies on $\overline{WZ}$ and segment $\overline{OX}$ bisects $\angle YOW$. What is the measure of $\angle XOW$?

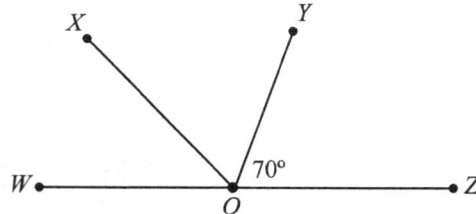

F. 40°
G. 55°
H. 65°
J. 85°
K. 110°

Lines and Angles
Geometry Problem Set 18

5. In the figure below, lines k and l are NOT parallel. If $a = 62$, which of the following CANNOT be the value of b?

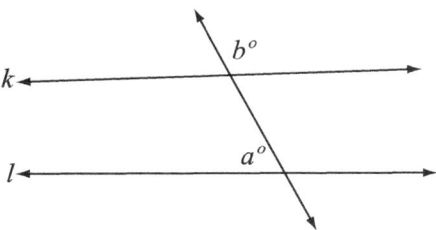

   A. 117
   B. 118
   C. 119
   D. 120
   E. 121

6. In the figure below, three lines intersect at a point. If $x = 80$ and $t = 15$, what is the value of w?

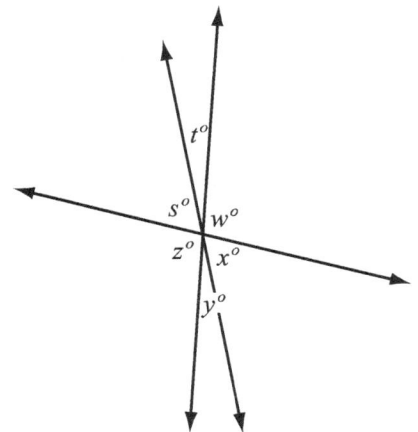

   F. 75
   G. 80
   H. 85
   J. 95
   K. 100

Lines and Angles
Geometry Problem Set 18

7. In the figure below, line m is parallel to line l. If $b = 21$, what is the value of a?

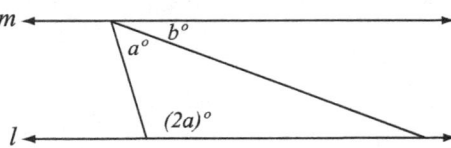

   A. 51
   B. 52
   C. 53
   D. 54
   E. 55

8. Three lines are drawn in a plane such that there is exactly one point of intersection, where all of the lines intersect. Into how many non-overlapping regions do these lines divide the plane?

   F. 3
   G. 4
   H. 5
   J. 6
   K. 7

Lines and Angles
Geometry Problem Set 18

9. In the figure below, which of the following is greatest?

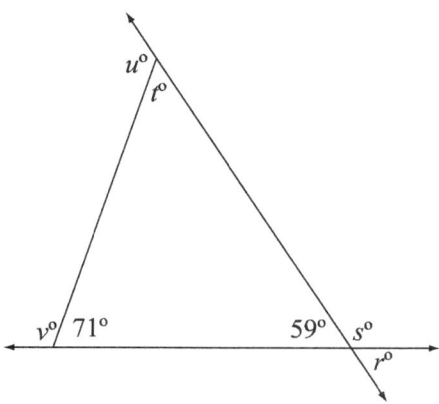

- A. r
- B. s
- C. t
- D. u
- E. v

10. In the figure below, what is the value of $a + b + c + d$ in terms of x?

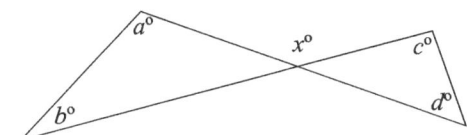

- F. $\dfrac{x}{4}$
- G. $\dfrac{x}{2}$
- H. x
- J. $2x$
- K. $4x$

Lines and Angles
Geometry Problem Set 18

11. In the figure below, ∠BAC measures 40°, ∠ABC measures 75°, and points B, C, and D are collinear. What is the measure of ∠ACD?

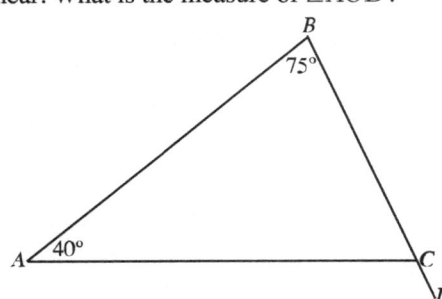

A. 65°
B. 75°
C. 105°
D. 115°
E. 140°

12. In the figure below, △NOP is an isosceles triangle, where $\overline{NO} = \overline{NP}$. $\overline{PA}$ intersects $\overline{NB}$ at O. What is the measure of ∠N?

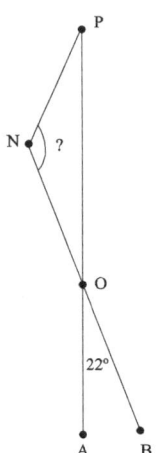

F. 22°
G. 44°
H. 46°
J. 136°
K. 158°

Lines and Angles
Geometry Problem Set 18

13. Two distinct line segments, $\overline{AB}$ and $\overline{CD}$ intersect so that point A lies on $\overline{CD}$. Which of the following statements must be true about the two angles this intersection forms?

- **A.** The difference between $\angle CAB$ and $\angle BAD$ is less than $45°$
- **B.** The difference between $\angle CAB$ and $\angle BAD$ is $180°$
- **C.** $\angle CAB$ and $\angle BAD$ are both $90°$
- **D.** The sum of $\angle CAB$ and $\angle BAD$ is $90°$
- **E.** The sum of $\angle CAB$ and $\angle BAD$ is $180°$

14. In the figure below, $\overline{JK}$ is parallel to $\overline{MN}$, the measure of $\angle JKL$ is $31°$, and the measure of $\angle KLM$ is $79°$. What is the measure of $\angle LMN$?

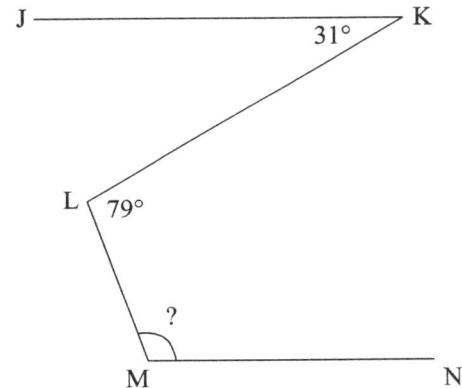

- **F.** $48°$
- **G.** $70°$
- **H.** $101°$
- **J.** $110°$
- **K.** $132°$

Lines and Angles
Geometry Problem Set 18

15. The circumference of circle C, shown below, is 18. Arc $CD = 3$. What is the value of x?

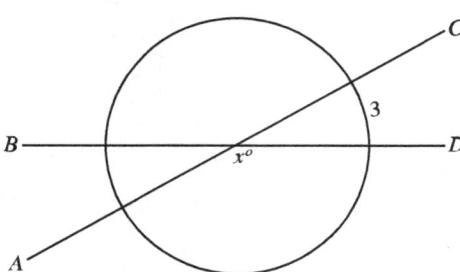

- A. 30
- B. 60
- C. 100
- D. 120
- E. 160

16. In the figure below, lines i and j intersect each other at point A. What is the measure of the largest angle in $\triangle ABC$?

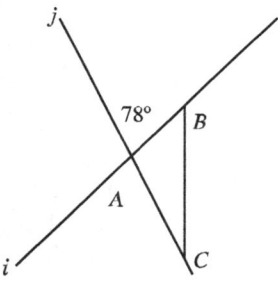

- F. 39°
- G. 78°
- H. 88°
- J. 102°
- K. 141°

Lines and Angles
Geometry Problem Set 18

17. In the figure below, what is the value of x?

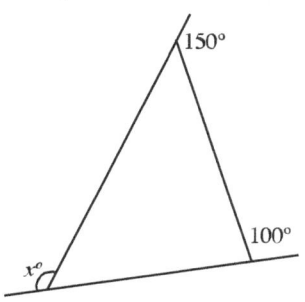

- **A.** 30
- **B.** 80
- **C.** 100
- **D.** 110
- **E.** 350

18. In the figure below, $\triangle ABC$, $\overline{AB} = \overline{BC} = \overline{AC} = 4$. What is the value of x?

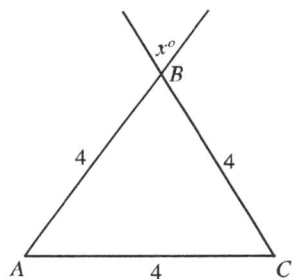

- **F.** 45
- **G.** 60
- **H.** 75
- **J.** 78
- **K.** 85

Lines and Angles
Geometry Problem Set 18

19. In the figure below, what is the value of b?

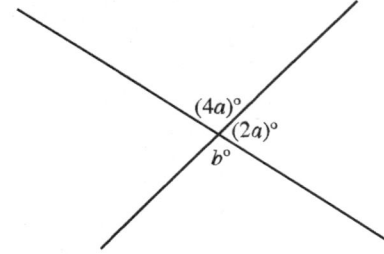

- **A.** 30
- **B.** 60
- **C.** 100
- **D.** 120
- **E.** 150

20. In the figure below, there is a rhombus $ABCD$. What is the measure of $\angle ADC$?

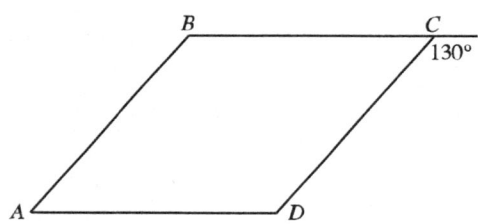

- **F.** 50°
- **G.** 85°
- **H.** 105°
- **J.** 120°
- **K.** 130°

Lines and Angles
Geometry Problem Set 18

21. In the figure below, what is the value of x in trapezoid $ABCD$?

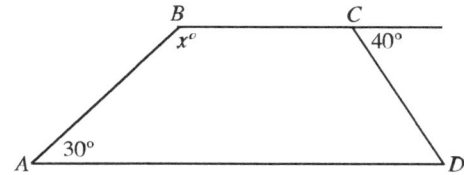

- **A.** 120
- **B.** 130
- **C.** 150
- **D.** 160
- **E.** 170

22. In the figure below, what is the value of a?

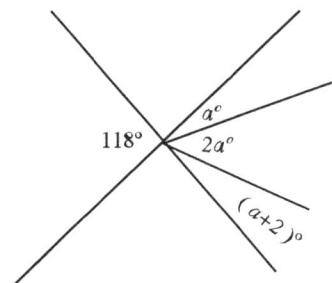

- **F.** 18
- **G.** 25
- **H.** 29
- **J.** 31
- **K.** 33

Lines and Angles
Geometry Problem Set 18

23. In the figure below, what is the value of x?

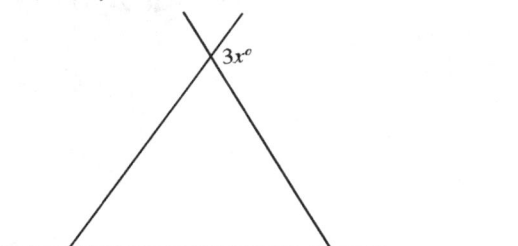

- **A.** 35
- **B.** 40
- **C.** 50
- **D.** 65
- **E.** 80

24. There is a polygon with 6 sides. What is the sum of the exterior angles?

- **F.** 180°
- **G.** 360°
- **H.** 540°
- **J.** 720°
- **K.** 900°

25. In the figure below, what is the value of b?

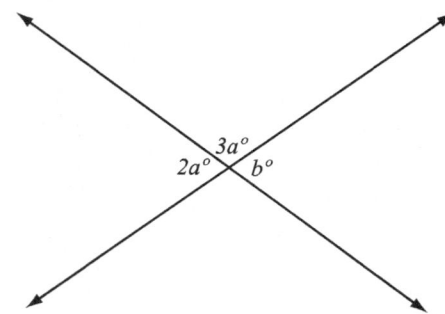

- **A.** 36
- **B.** 60
- **C.** 72
- **D.** 84
- **E.** 108

Lines and Angles
Geometry Problem Set 18

Answer Key

#	Answer	Frequency	Difficulty
1	B	popular	1
2	G	popular	2
3	C	popular	1
4	G	popular	1
5	B	popular	2
6	H	popular	1
7	C	popular	2
8	J	rare	2
9	D	average	2
10	J	average	4
11	D	average	1
12	J	popular	3
13	E	popular	1
14	K	popular	1
15	D	average	2
16	J	average	2
17	D	popular	2
18	G	popular	1
19	D	popular	2
20	K	popular	2
21	C	popular	1
22	H	popular	2
23	C	average	3
24	G	popular	3
25	C	popular	3

Pythagorean Theorem
Geometry Problem Set 19

1. In the figure below, what is the area of square $ABCD$?

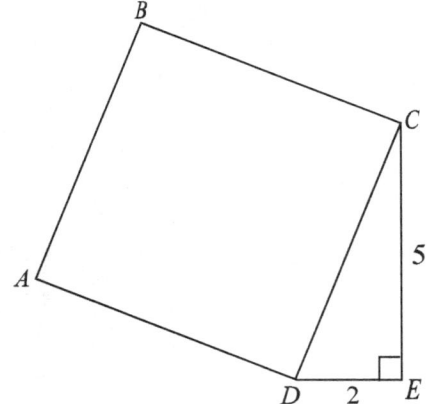

   A. 21
   B. 25
   C. 29
   D. 35
   E. 49

   DO YOUR FIGURING HERE

2. In $\triangle ABC$ below, $\overline{AD} = 6$, $\overline{DC} = 3$, and $\overline{AB} = 10$. What is the area of the $\triangle CBD$?

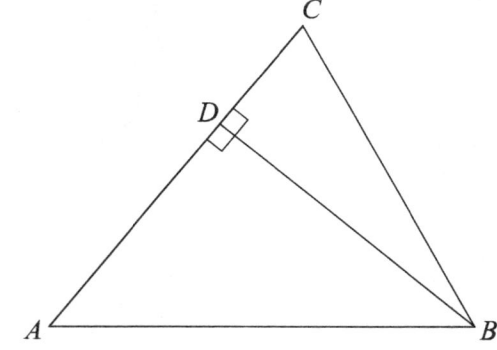

   F. 12
   G. 15
   H. 29
   J. 30
   K. 90

Pythagorean Theorem
Geometry Problem Set 19

3. In the figure below, if *VWXY* is a square, what is the area of polygon *VWXYZ*?

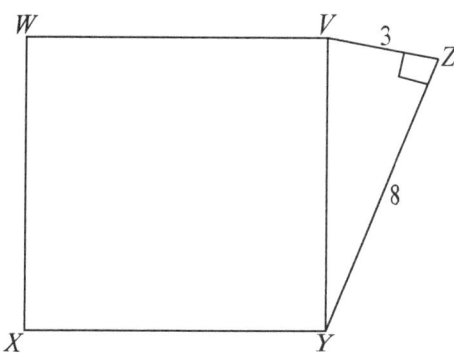

- A. 85
- B. 88
- C. 97
- D. 121
- E. 144

4. The figure below is a right triangle. What is the value of $16 + a^2$?

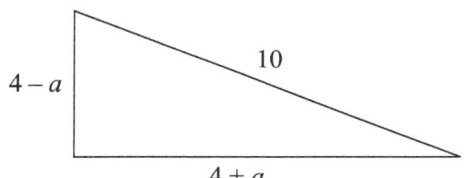

- F. 22
- G. 50
- H. 65
- J. 68
- K. 100

Pythagorean Theorem
Geometry Problem Set 19

5. In $\triangle QRS$ below, $\overline{QT} = 12$, $\overline{QR} = 13$, and $\overline{TS} = \sqrt{39}$. What is the length of $\overline{RS}$?

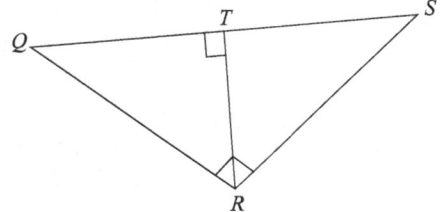
Note: Figure not drawn to scale

A. $2\sqrt{2}$

B. 8

C. 12

D. 16

E. 24

DO YOUR FIGURING HERE

Pythagorean Theorem
Geometry Problem Set 19

Answer Key

#	Answer	Frequency	Difficulty
1	C	average	2
2	F	average	4
3	A	average	3
4	G	average	4
5	B	average	2

Triangle Inequality Theorem
Geometry Problem Set 20

1. Which of the following could be the lengths of the sides of a triangle?

    A. 1, 4, 5
    B. 2, 2, 4
    C. 3, 5, 8
    D. 4, 7, 13
    E. 5, 6, 10

2. If a is an integer and $3 < a < 10$, how many different triangles are possible with sides measuring 3, 10, and a?

    F. 2
    G. 4
    H. 5
    J. 6
    K. 8

3. The lengths of two sides of a triangle are 11 and 3, respectively. If the third side must be an integer, what is one possible length of the third side?

DO YOUR FIGURING HERE

Triangle Inequality Theorem
Geometry Problem Set 20

4. In the figure below, if the legs of △QRS are parallel to the axes, which of the following could be the lengths of the sides of △QRS?

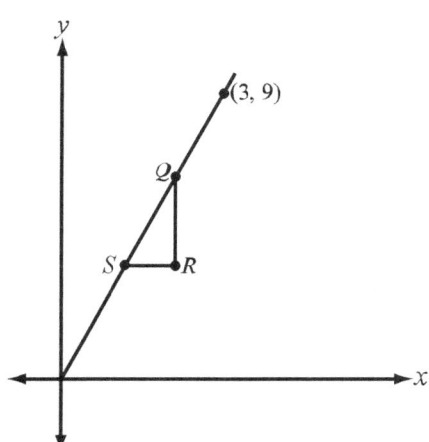

F. 1, 3, 5
G. 1, 3, √10
H. 3, 4, 5
J. 3, 6, √45
K. 3, 3, √2

DO YOUR FIGURING HERE

Triangle Inequality Theorem
Geometry Problem Set 20

Answer Key

#	Answer	Frequency	Difficulty
1	E	average	2
2	F	average	3
3	9, 10, 11, 12, 13	average	2
4	G	average	4

Similar Triangles
Geometry Problem Set 21

1. In the figure below, what is the value of $\dfrac{AE}{AD}$?

   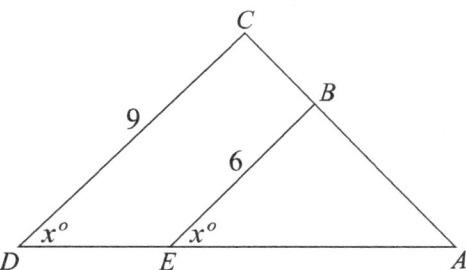

   A. $\dfrac{1}{9}$

   B. $\dfrac{2}{9}$

   C. $\dfrac{4}{9}$

   D. $\dfrac{2}{3}$

   E. $\dfrac{5}{3}$

2. In the figure below, $\triangle ABC$ is similar to $\triangle DEF$. What is the value of x?

   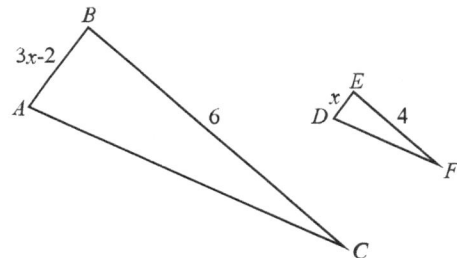

   F. $\dfrac{1}{3}$

   G. $\dfrac{3}{4}$

   H. $\dfrac{6}{7}$

   J. 1

   K. $\dfrac{4}{3}$

Similar Triangles
Geometry Problem Set 21

3. In the figure below, point B bisects segment $\overline{AC}$ and point E bisects segment $\overline{AD}$. What is the ratio of segment $\overline{BE}$ to segment $\overline{CD}$?

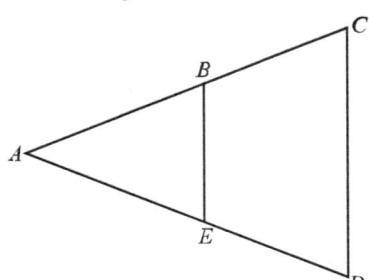

DO YOUR FIGURING HERE

A. 1:2
B. 1:3
C. 1:4
D. 1:5
E. 1:8

4. In the figure below, $\triangle NOP$ is similar to $\triangle RST$. What is the length of side $\overline{NP}$?

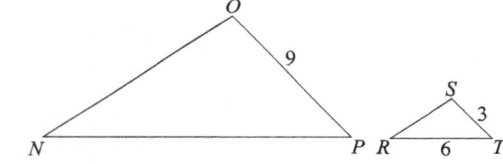

F. 18
G. 19.5
H. 21
J. 22.5
K. 24

Similar Triangles
Geometry Problem Set 21

5. In the figure below, △PQR and △RST are equilateral and line segment $\overline{PT}$ has a length of 30. What is the sum of the perimeters of the two triangles?

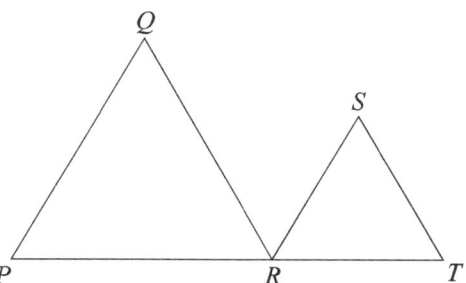

6. Cassidy is making a map of her route from her home, A, to her school, B, and her friend's home, C. She shows the 2-mile distance between home and school as 13 centimeters on the map. The distance between Cassidy's school and her friend's home is 1.5 miles. How long should the line between her school and her friend's home be on the map? Round to the nearest hundredth of a centimeter.

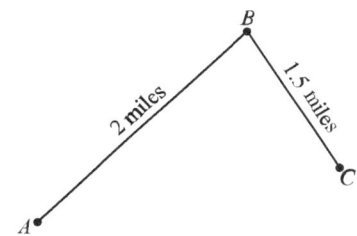

F. 5.34 cm
G. 7.89 cm
H. 8.67 cm
J. 9.75 cm
K. 17.33 cm

Similar Triangles
Geometry Problem Set 21

7. In the parallelogram $ABCD$, shown below, which of the following must be true?

   I. $\triangle ABE$ is similar to $\triangle BCE$
   II. $\angle BCE = \angle BAE$
   III. $\overline{AE} = \overline{EC}$

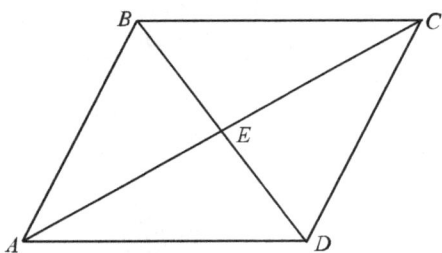

   A. I only
   B. II only
   C. III only
   D. I and III only
   E. I, II, and III

8. In the picture below, the length of $\overline{AF}$ is 8 and $\overline{AE}$ is 14. If $\overline{FC}$ is 12, then what is the length of $\overline{ED}$?

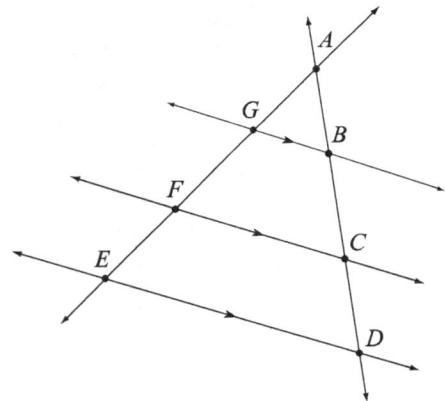

   F. 16
   G. 18
   H. 20
   J. 21
   K. 24

Similar Triangles
Geometry Problem Set 21

9. In the figure below, △ABC is similar to △DEF. The area of △ABC is 36. What is the area of △DEF? Round to the nearest integer.

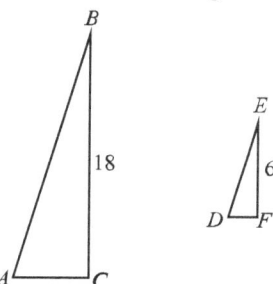

- **A.** 4
- **B.** 8
- **C.** 10
- **D.** 12
- **E.** 18

10. Given the 2 similar right triangles shown below with dimensions given in centimeters, what is the area, in square centimeters, of the larger triangle?

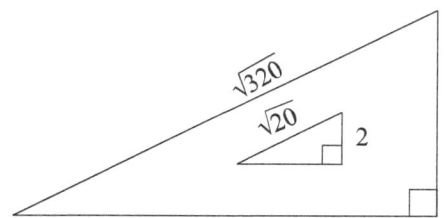

- **F.** 4
- **G.** 8
- **H.** 16
- **J.** 64
- **K.** 128

Similar Triangles
Geometry Problem Set 21

11. For two triangles, $\triangle ABC$ and $\triangle DEF$, $\overline{AB} = 2\overline{DE}$, $\overline{BC} = 2\overline{EF}$, and $\angle CAB = \angle FDE$. Given this information, which of the following statements is true?

    A. $\angle ABC = \angle DEF$

    B. $\angle BCA = \angle EFD$

    C. $\overline{AC} = 2\overline{DF}$

    D. All of the above

    E. None of the above

DO YOUR FIGURING HERE

Similar Triangles
Geometry Problem Set 21

Answer Key

#	Answer	Frequency	Difficulty
1	D	average	2
2	K	average	2
3	A	average	1
4	F	average	3
5	90	average	2
6	J	average	2
7	C	popular	2
8	J	average	3
9	A	average	2
10	J	average	3
11	E	average	3

Isosceles Triangles
Geometry Problem Set 22

1. In $\triangle PQR$, shown in the figure, $\overline{PR} = \overline{QR}$. Which of the following must be true?

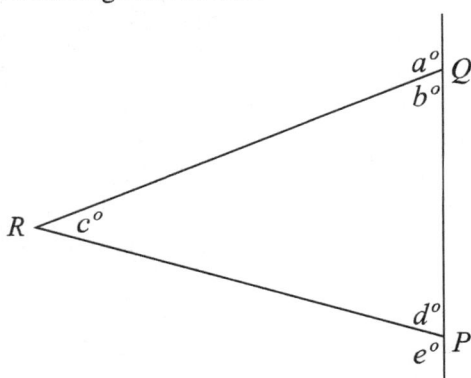

A. $a = b$
B. $c = e$
C. $e = a$
D. $a = c$
E. $d = c$

2. In the figure, Q is located at the origin and $\overline{QR} = \overline{QS}$. What is the value of y at point S?

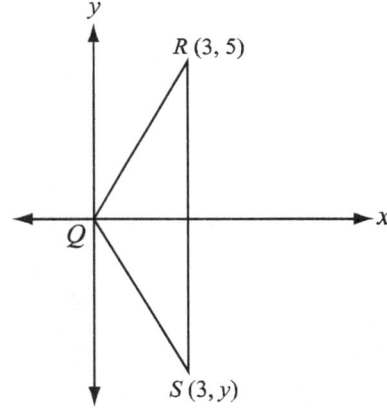

F. -5
G. -3
H. 0
J. 3
K. 5

Isosceles Triangles
Geometry Problem Set 22

3. The triangle shown in the figure is isosceles and $\overline{PQ} > \overline{PR}$. Which of the following must be FALSE?

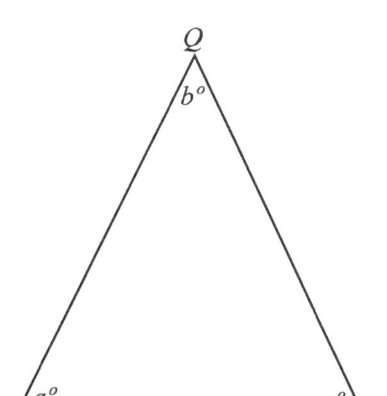

- A. $a = b$
- B. $b = c$
- C. $a = c$
- D. $\overline{PQ} > \overline{QR}$
- E. $\overline{PQ} = \overline{QR}$

4. In right triangle ABC, shown in the figure, line $\overline{AC}$ = line $\overline{BC}$ = 3. What is the value of x?

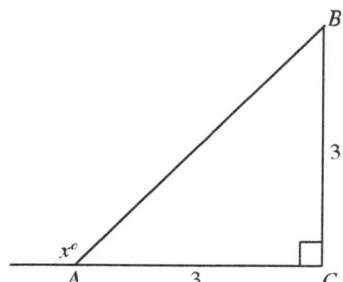

- F. 45
- G. 60
- H. 120
- J. 135
- K. 150

Isosceles Triangles
Geometry Problem Set 22

5. In the figure, $\triangle ABC$ is an isosceles triangle. What is the measure of $\angle ABC$?

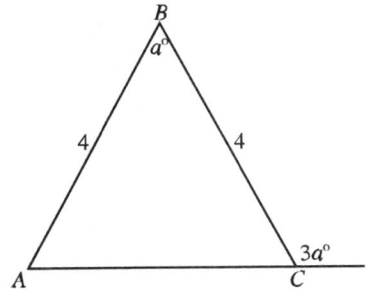

A. $28°$

B. $30°$

C. $36°$

D. $60°$

E. $108°$

6. In the figure, $\overline{AB} = \overline{BC}$ and $\angle BAC = 39°$. What is the value of x?

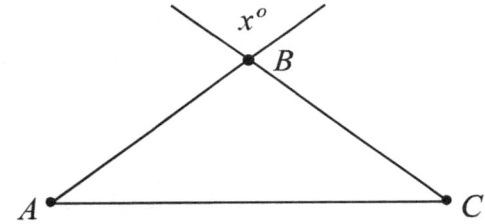

Isosceles Triangles
Geometry Problem Set 22

Answer Key

#	Answer	Frequency	Difficulty
1	C	popular	2
2	F	popular	1
3	B	popular	2
4	J	popular	2
5	C	popular	3
6	102°	popular	2

Triangle Area
Geometry Problem Set 23

1. What is the area of the △ABC shown below?

   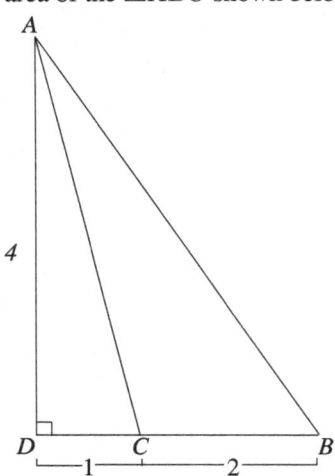

   A. 4
   B. 6
   C. 8
   D. 10
   E. 12

2. In the figure below, $\overline{AB} = \overline{BC} = 13$ and $\overline{AC} = 10$. What is the area of △ABC?

   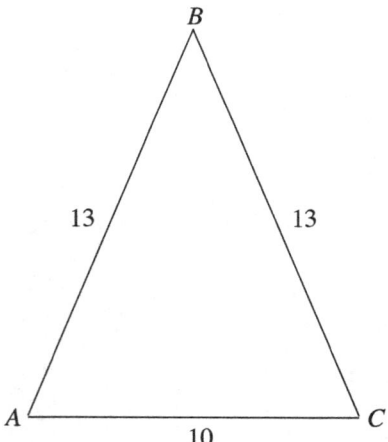

   F. 30
   G. 32.5
   H. 48
   J. 60
   K. 65

Triangle Area
Geometry Problem Set 23

3. In the xy-coordinate plane, $\triangle ABC$ has points $A\,(-2,-4)$, $B\,(-2,6)$, and $C\,(2,6)$. What is the area of $\triangle ABC$?

   A. 4
   B. 8
   C. 12
   D. 20
   E. 40

4. If a triangle has an area of 12 and a height of 6, then what is the measure of the base?

   F. 2
   G. 3
   H. 4
   J. 5
   K. 8

5. In the figure below, right triangle ABC is half of the area of the square $ABCD$, which has a perimeter of 20. What is the area of $\triangle ABC$?

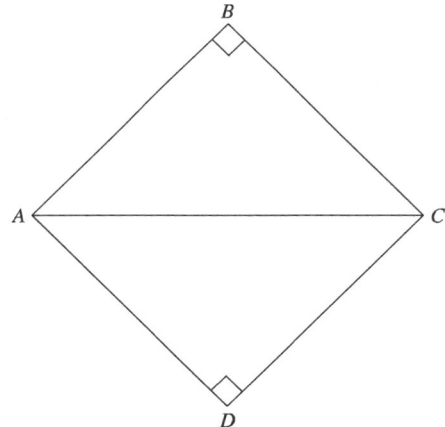

   A. 6.25
   B. 10
   C. 11.5
   D. 12.5
   E. 25

Triangle Area
Geometry Problem Set 23

6. In the figure below, the equilateral triangle has a side $\overline{AB} = 4$. What is the area of $\triangle ABC$, to the thousandth place?

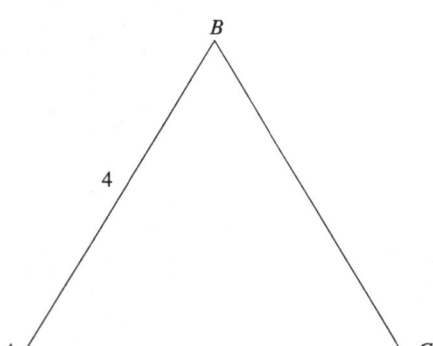

- **F.** 3.464
- **G.** 4.000
- **H.** 5.657
- **J.** 6.928
- **K.** 8.125

7. In the figure below, the hypotenuse of right triangle ACD forms the base of right triangle ABC. $\overline{AC} = 5$, $\overline{BC} = 10$, and $\overline{CD} = 4$. What is the area of quadrilateral $ABCD$?

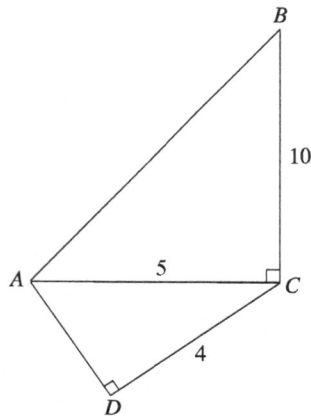

- **A.** 12
- **B.** 18
- **C.** 30
- **D.** 31
- **E.** 40

Triangle Area
Geometry Problem Set 23

8. In the figure below, $\overline{AC} = 12$, $\overline{BC} = 10$, and $\angle ABC = 30°$. What is the area of $\triangle ABC$?

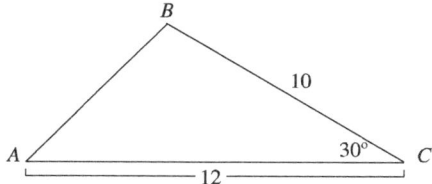

   F. 25
   G. 30
   H. 36
   J. 50
   K. 60

9. In the figure below, $\overline{AB} = \overline{AC} = \overline{BC} = 6$. What is the area of $\triangle ABC$ to the nearest thousandth place?

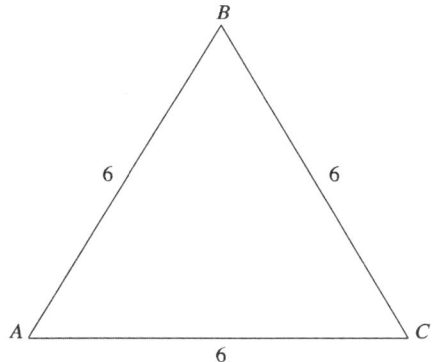

   A. 7.794
   B. 10.089
   C. 15.588
   D. 18.000
   E. 31.177

Triangle Area
Geometry Problem Set 23

10. In the figure below, $\overline{AD} = x$ and $\overline{BD} = x+1$. The area of $\triangle ABC$ is 6. What is the value of x?

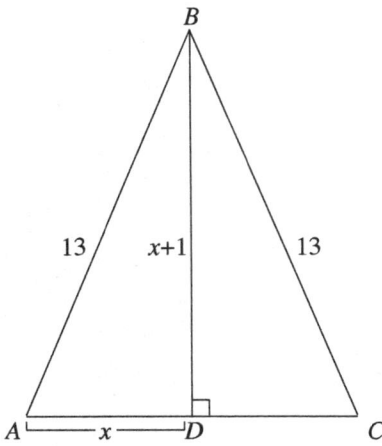

- **F.** 1
- **G.** 2
- **H.** 3
- **J.** 5
- **K.** 6

11. In the figure below, $\triangle ABC$ is inscribed in the rectangle $ABCD$. If the area of $\triangle ABC$ is 5, then what is the perimeter of rectangle $ABCD$?

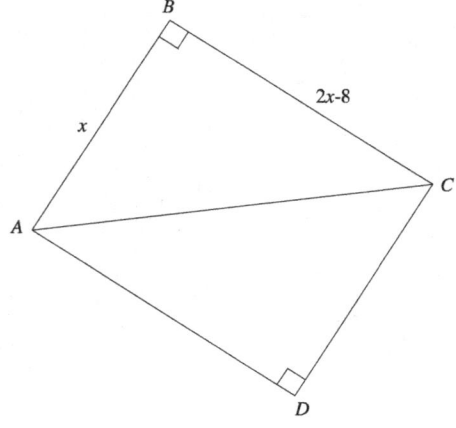

- **A.** 3
- **B.** 5
- **C.** 10
- **D.** 12
- **E.** 14

Triangle Area
Geometry Problem Set 23

12. What is the greatest possible area of a triangle with one side of length 7 and another side of length 12?

    F. 21
    G. 35
    H. 42
    J. 84
    K. 168

13. △ABC, shown below, is entirely contained within a circle, O. If $\overline{AC} = 3$ and the area of $ABC = 6$, then what is the minimum area of circle O?

    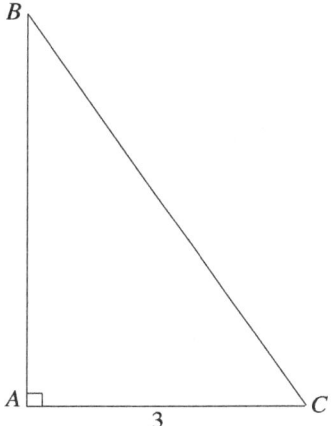

    A. 12
    B. 4π
    C. 6.25π
    D. 13.5π
    E. 25π

Triangle Area
Geometry Problem Set 23

14. In the figure below, △ABC is similar to △DEF. The area of △ABC is 36. What is the area of △DEF? Round to the nearest integer.

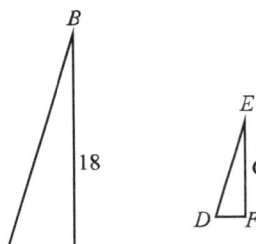

- **F.** 4
- **G.** 8
- **H.** 10
- **J.** 12
- **K.** 18

15. In the figure below, the area of △BCD is 15, $\overline{AB} = 2x$, $\overline{AD} = x$, and $\overline{CD} = x + 2$. What is the value of x?

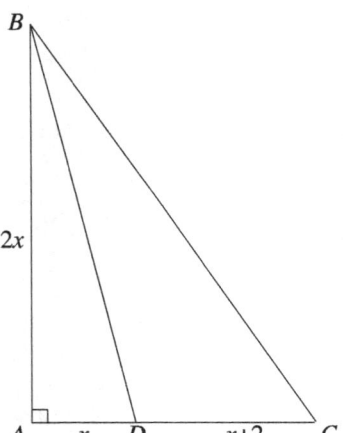

- **A.** 2
- **B.** 3
- **C.** 5
- **D.** 6
- **E.** 10

Triangle Area
Geometry Problem Set 23

Answer Key

#	Answer	Frequency	Difficulty
1	A	popular	1
2	J	popular	2
3	D	popular	1
4	H	popular	1
5	D	popular	2
6	J	popular	2
7	D	popular	2
8	G	popular	3
9	C	popular	3
10	G	popular	3
11	E	popular	3
12	H	popular	3
13	C	popular	3
14	F	average	2
15	B	popular	4

Special Right Triangles
Quick Drill

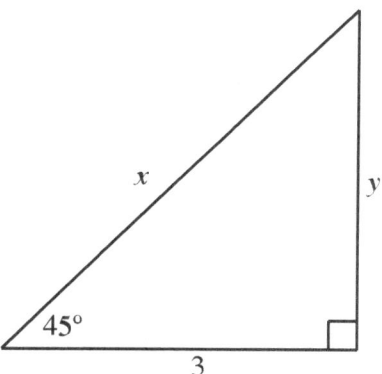

1. What are the values for x and y? (Leave your answers in simplest form with radicals.)

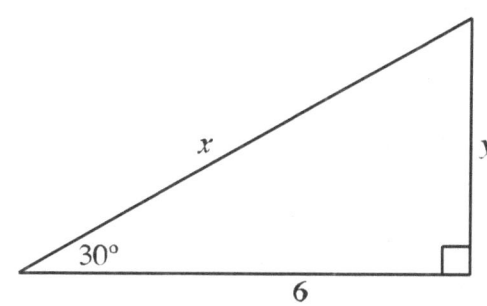

3. What are the values for x and y? (Leave your answers in simplest form with radicals.)

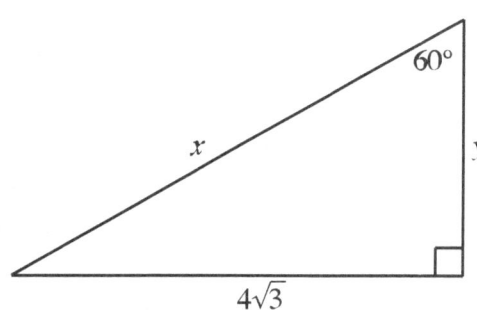

2. What are the values for x and y? (Leave your answers in simplest form with radicals.)

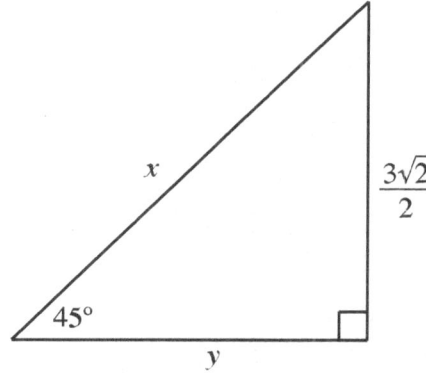

4. What are the values for x and y? (Leave your answers in simplest form with radicals.)

Special Right Triangles
Quick Drill

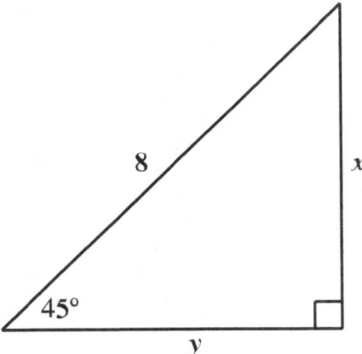

5. What are the values for x and y? (Leave your answers in simplest form with radicals.)

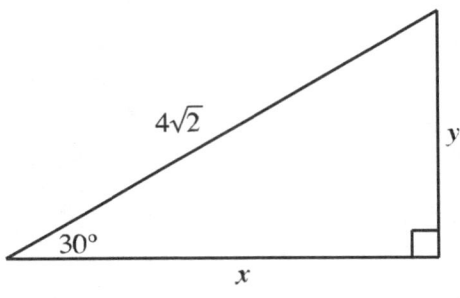

7. What are the values for x and y? (Leave your answers in simplest form with radicals.)

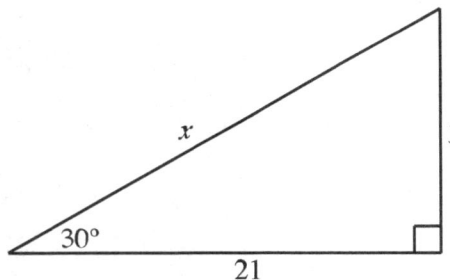

6. What are the values for x and y? (Leave your answers in simplest form with radicals.)

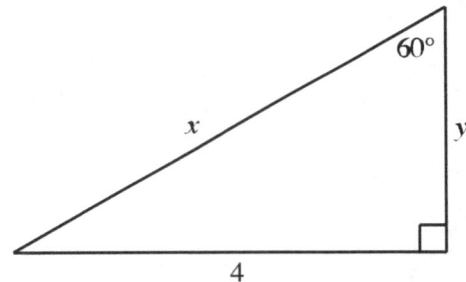

8. What are the values for x and y? (Leave your answers in simplest form with radicals.)

Special Right Triangles
Quick Drill

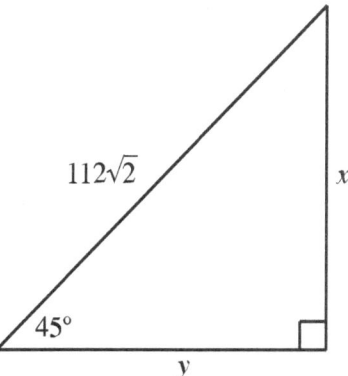

9. What are the values for x and y? (Leave your answers in simplest form with radicals.)

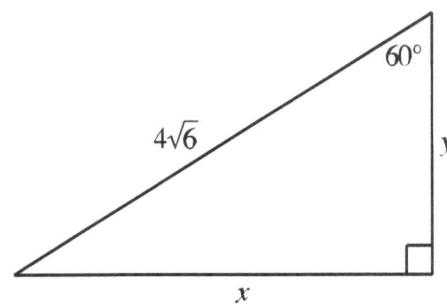

10. What are the values for x and y? (Leave your answers in simplest form with radicals.)

Special Right Triangles
Quick Drill

Answer Key

#	Answer
1	$x = 3\sqrt{2}$ and $y = 3$
2	$x = 8$ and $y = 4$
3	$x = 4\sqrt{3}$ and $y = 2\sqrt{3}$
4	$x = 3$ and $y = \dfrac{3\sqrt{2}}{2}$
5	$x = 4\sqrt{2}$ and $y = 4\sqrt{2}$
6	$x = 14\sqrt{3}$ and $y = 7\sqrt{3}$
7	$x = 2\sqrt{6}$ and $y = 2\sqrt{2}$
8	$x = \dfrac{8\sqrt{3}}{3}$ and $y = \dfrac{4\sqrt{3}}{3}$
9	$x = 112$ and $y = 112$
10	$x = 6\sqrt{2}$ and $y = 2\sqrt{6}$

Squares and 45-45-90 Triangles
Geometry Problem Set 24

1. What is the area of square $ABCD$?

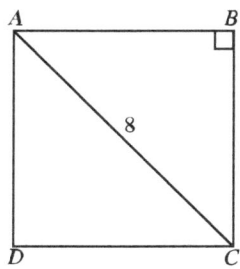

   A. 16
   B. $4\sqrt{2}$
   C. $16\sqrt{2}$
   D. 32
   E. $8\sqrt{2}$

2. What is the perimeter of quadrilateral $WXYZ$?

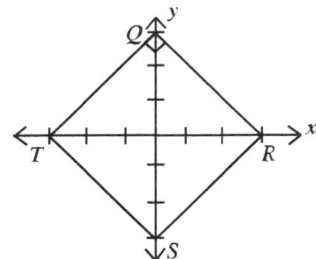

   F. $3\sqrt{2}$
   G. $12\sqrt{2}$
   H. 18
   J. 12
   K. 6

Squares and 45-45-90 Triangles
Geometry Problem Set 24

3. In the figure below, what is the perimeter of square $EFGH$?

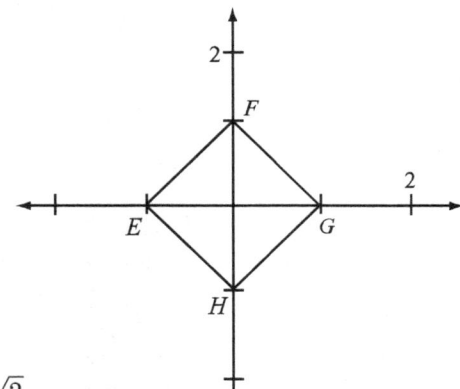

A. $\sqrt{2}$
B. $4\sqrt{2}$
C. $4\sqrt{3}$
D. $8\sqrt{2}$
E. $8\sqrt{3}$

DO YOUR FIGURING HERE

4. If the $\triangle ABC$ (shown below) is isosceles, then what is the total area of the triangle?

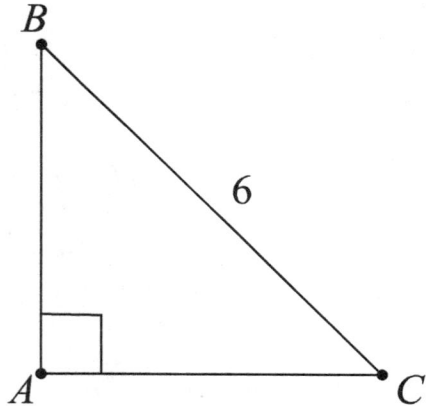

F. $6\sqrt{2}$
G. 9
H. $9\sqrt{2}$
J. 12
K. 18

Squares and 45-45-90 Triangles
Geometry Problem Set 24

5. On the staircase below, both the depth and the height of each step are x, and each step forms a right angle. What is the value of z in terms of x?

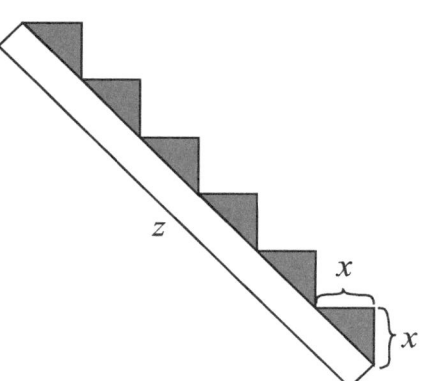

- A. $6x\sqrt{2}$
- B. $12x\sqrt{2}$
- C. $6x\sqrt{3}$
- D. $6x$
- E. 12

6. The quadrilateral $QRST$, shown below, is a square. What is the total area of $QRST$?

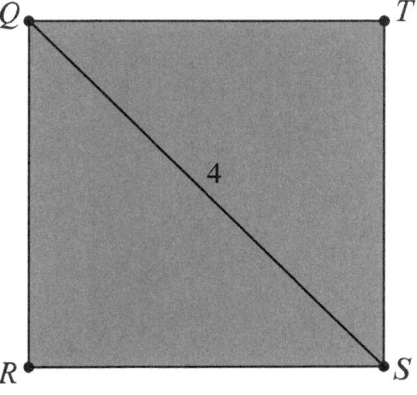

Squares and 45-45-90 Triangles
Geometry Problem Set 24

Answer Key

#	Answer	Frequency	Difficulty
1	D	popular	1
2	G	popular	1
3	B	popular	2
4	G	popular	3
5	A	popular	3
6	8	popular	3

Equilateral and 30-60-90 Triangles
Geometry Problem Set 25

1. In $\triangle PRS$ below, $\overline{PR} = 4$, and T is the midpoint of $\overline{SP}$. What is the length of $\overline{RS}$?

   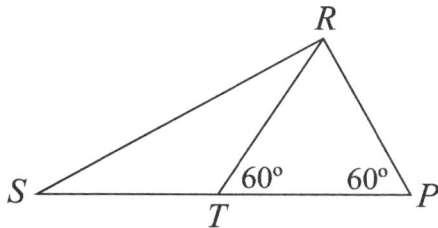

   DO YOUR FIGURING HERE

   A. $4\sqrt{2}$
   B. $4\sqrt{3}$
   C. $6\sqrt{2}$
   D. $6\sqrt{3}$
   E. $8\sqrt{2}$

2. In $\triangle WXY$ below, sides $\overline{WX}$ and $\overline{XY}$ are equal in length and the value of a is 30. What is the value of b?

   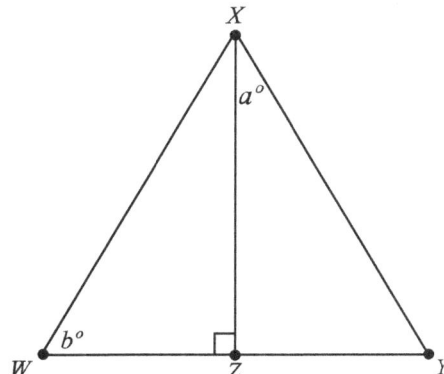

   F. 40
   G. 45
   H. 50
   J. 55
   K. 60

Equilateral and 30-60-90 Triangles
Geometry Problem Set 25

3. In the figure below, each of the four small triangles is equilateral and has a perimeter of 36. What is the perimeter of △CAT?

DO YOUR FIGURING HERE

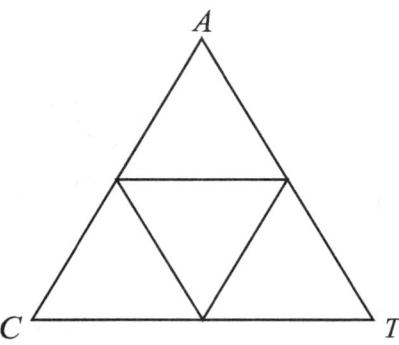

- A. 36
- B. 48
- C. 56
- D. 72
- E. 108

4. What is the perimeter of the triangle shown below?

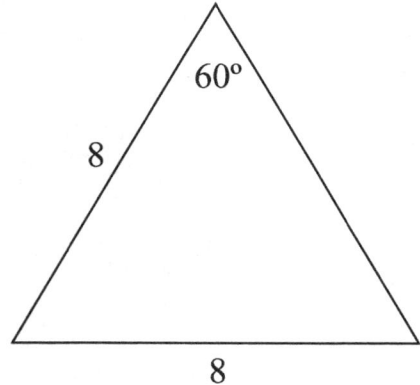

- F. 8
- G. 16
- H. $4\sqrt{3}$
- J. 24
- K. 64

Equilateral and 30-60-90 Triangles
Geometry Problem Set 25

5. In the figure below, $ABCD$ is a square and $\triangle BCE$ is an equilateral triangle. If $ABCD$ has an area of 4, what is the area of $\triangle BCE$?

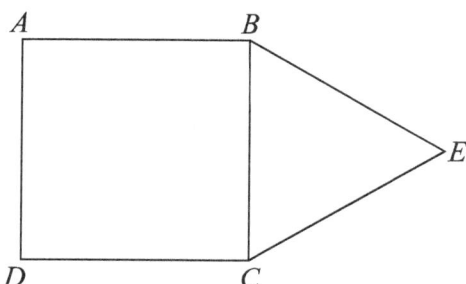

DO YOUR FIGURING HERE

A. 1
B. 2
C. 3
D. $\sqrt{3}$
E. $\sqrt{2}$

6. In the figure below, the length of $\overline{AC}$ (not shown) is 8 and $\overline{AB} = \overline{BC}$. What is the total area of quadrilateral $ABCD$?

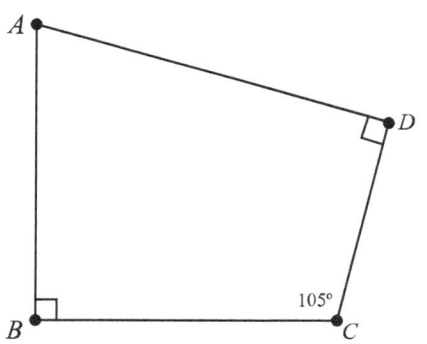

F. 32
G. $8\sqrt{2} + 16\sqrt{3}$
H. $16 + 16\sqrt{3}$
J. $16 + 8\sqrt{3}$
K. $16\sqrt{2} + 16$

Equilateral and 30-60-90 Triangles
Geometry Problem Set 25

7. If the equilateral triangle below has a height of 12, what is the perimeter?

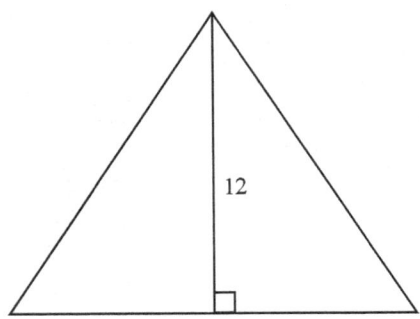

- A. 24
- B. 36
- C. $12\sqrt{3}$
- D. $24\sqrt{3}$
- E. $36\sqrt{3}$

DO YOUR FIGURING HERE

8. In $\triangle ABC$ below, the measure of $\angle BAC$ is 30° and the measure of $\angle ACB$ is 60°. What is the length of segment AC?

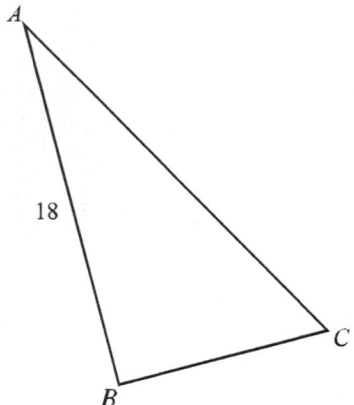

- F. 6
- G. 9
- H. $6\sqrt{3}$
- J. 12
- K. $12\sqrt{3}$

Equilateral and 30-60-90 Triangles
Geometry Problem Set 25

Answer Key

#	Answer	Frequency	Difficulty
1	B	popular	3
2	K	popular	1
3	D	popular	2
4	J	popular	1
5	D	popular	3
6	J	popular	3
7	D	popular	2
8	K	popular	3

Polygons
Geometry Problem Set 26

1. The figure shown below is a regular octagon. What is the measure of x?

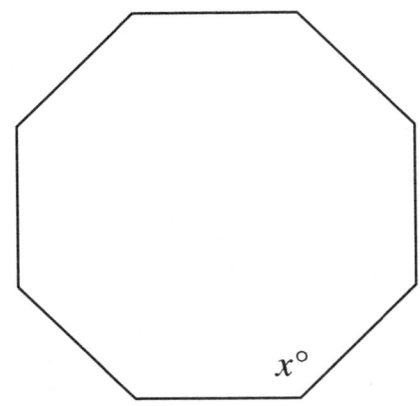

2. If the figure shown below is a regular polygon, what is the measure of b?

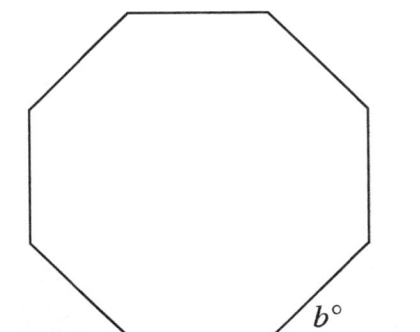

   F. 30
   G. 45
   H. 60
   J. 120
   K. 134

Polygons
Geometry Problem Set 26

3. If point O is the center of the regular hexagon drawn below, what is the measure of $\angle AOF$?

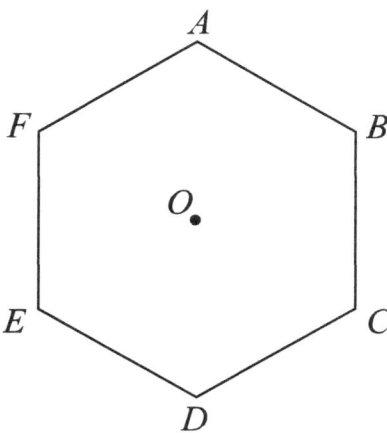

- A. 40°
- B. 45°
- C. 50°
- D. 60°
- E. 72°

4. In the figure below, $PQRSTUVW$ is a regular octagon. If point O is located at the center, then what is the value of a?

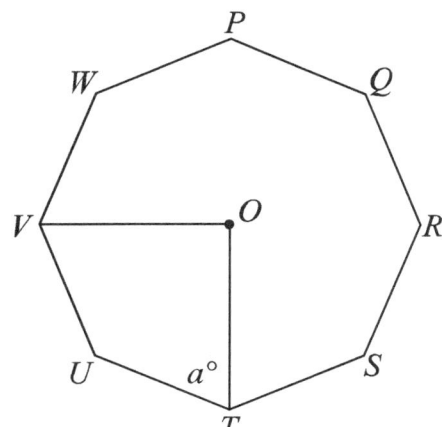

- F. 40
- G. 60
- H. 67.5
- J. 70.5
- K. 72.5

Polygons
Geometry Problem Set 26

5. In the figure below, $FGHIJK$ is a regular hexagon. What is the value of x?

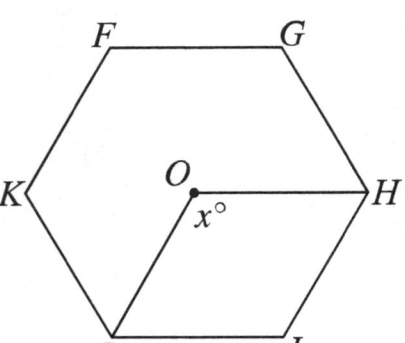

A. 60
B. 90
C. 108
D. 120
E. 135

6. Pentagon $PQRST$ in the figure below is a regular pentagon. The measure of $\angle PTQ$ is $36°$. If the measure of $\angle QTR$ is $x°$, then what is the value of x?

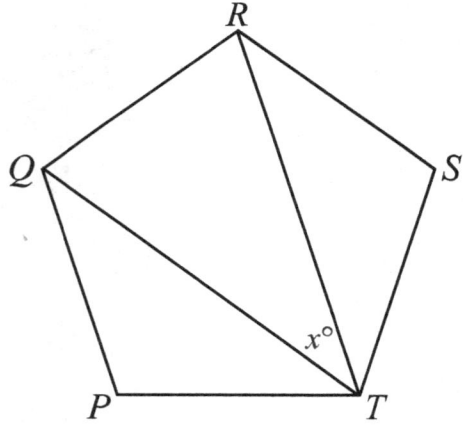

F. 36
G. 48
H. 56
J. 72
K. 108

Polygons
Geometry Problem Set 26

7. In the figure below, a polygon with equal angles and equal sides is partially covered by a gray piece of paper. If $a + b = 90$, how many sides does the polygon have?

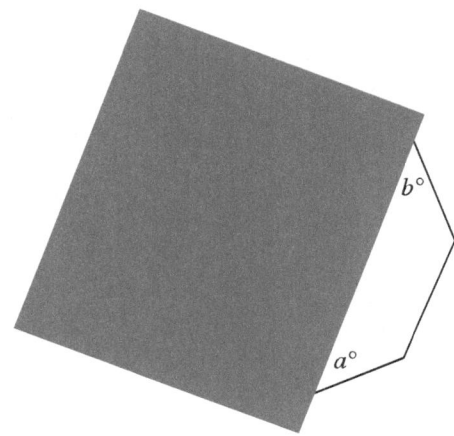

Note: Figure not drawn to scale

Polygons
Geometry Problem Set 26

Answer Key

#	Answer	Frequency	Difficulty
1	135	popular	1
2	G	popular	2
3	D	popular	3
4	H	popular	2
5	D	popular	2
6	F	popular	3
7	8	popular	4

Parallelograms
Geometry Problem Set 27

1. If the figure below is a parallelogram, what is the value of x?

   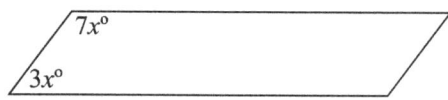

   A. 15
   B. 18
   C. 36
   D. 54
   E. 126

2. The difference between two consecutive angles in a parallelogram is $54°$. What is the measure of the smaller angle?

   F. $36°$
   G. $54°$
   H. $63°$
   J. $81°$
   K. $117°$

3. The figure below is a rhombus. All sides of a rhombus are the same length. If one diagonal is 12 centimeters and the other is 16 centimeters, how many centimeters long is each side of the rhombus?

   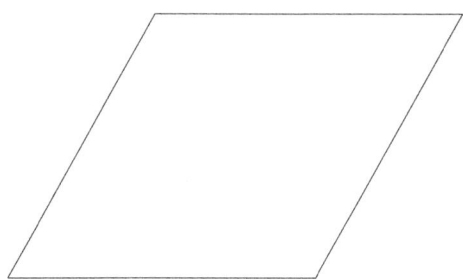

   A. $\sqrt{14}$
   B. $\sqrt{28}$
   C. 10
   D. $\sqrt{112}$
   E. 20

Parallelograms
Geometry Problem Set 27

4. The perimeter of a parallelogram is 70 inches, and the length of 1 side is 14 inches. If it can be determined, what are the lengths of the other 3 sides?

   F. 14, 14, 28
   G. 14, 14, 14
   H. 14, 21, 21
   J. 14, 28, 28
   K. Cannot be determined from the given information

5. All of the following statements concern parallelograms that are similar, congruent or both. Which statement is FALSE?

   A. Measures of corresponding angles of similar parallelograms are always equal.
   B. Measures of corresponding angles of congruent parallelograms are always equal.
   C. Lengths of corresponding sides of similar parallelograms are always equal.
   D. Lengths of corresponding sides of congruent parallelograms are always equal.
   E. Parallelograms that are congruent to each other are also always similar to each other.

DO YOUR FIGURING HERE

Parallelograms
Geometry Problem Set 27

6. In the figure below, there is a rhombus $ABCD$. What is the measure of $\angle ADC$?

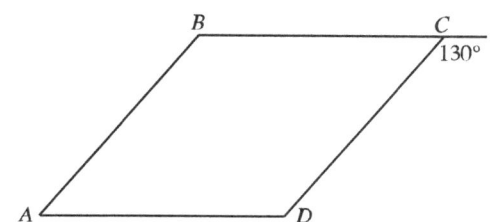

- F. $50°$
- G. $85°$
- H. $105°$
- J. $120°$
- K. $130°$

7. What is the value of x if quadrilateral $ABCD$, shown below, is a rhombus?

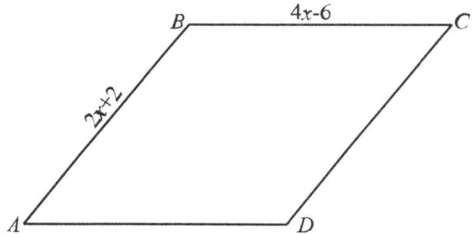

- A. $\dfrac{4}{3}$
- B. 2
- C. $\dfrac{8}{3}$
- D. 4
- E. $\dfrac{9}{2}$

Parallelograms
Geometry Problem Set 27

8. In the figure below, $\overline{AB} = 5$ and $\overline{AE} = 4$. If the area of parallelogram $ABCD$ is 15, then what is the measure of $\overline{BC}$?

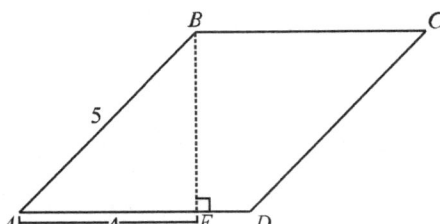

- F. 3
- G. $\dfrac{10}{3}$
- H. 5
- J. $\dfrac{22}{4}$
- K. 7

9. In the figure below, $\overline{BD} = 3x + 8$ and $\overline{BE} = 2x - 2$. What is the value of x?

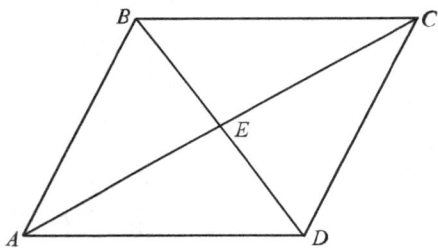

- A. 3
- B. 5
- C. 8
- D. 10
- E. 12

Parallelograms
Geometry Problem Set 27

10. In the parallelogram $ABCD$, shown below, which of the following must be true?

    I. $\triangle ABE$ is similar to $\triangle BCE$
    II. $\angle BCE = \angle BAE$
    III. $\overline{AE} = \overline{EC}$

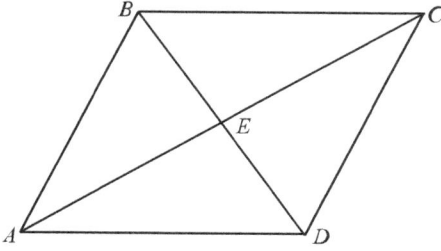

F. I only
G. II only
H. III only
J. I and III only
K. I, II, and III

11. A parallelogram is graphed in the xy-coordinate plane, with verticies $A(-2,3)$, $B(4,8)$, and $C(6,0)$. Which of the following are possible coordinates for the fourth vertex, point D?

A. $(-2,0)$
B. $(0,-5)$
C. $(1,-3)$
D. $(2,-6)$
E. Cannot be determined

Parallelograms
Geometry Problem Set 27

Answer Key

#	Answer	Frequency	Difficulty
1	B	popular	1
2	H	popular	2
3	C	popular	2
4	H	popular	1
5	C	average	1
6	K	popular	2
7	D	popular	1
8	H	popular	2
9	E	popular	2
10	H	popular	2
11	B	popular	3

Trapezoids
Geometry Problem Set 28

1. In the figure below, $\overline{AB}$ is parallel to $\overline{DC}$ and $ABCD$ is isosceles. What is the total area of quadrilateral $ABCD$?

   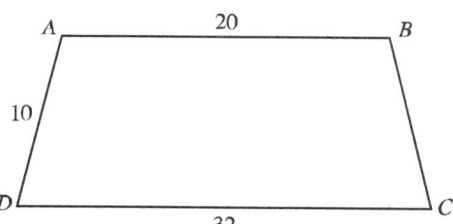

   A. 184
   B. 208
   C. 232
   D. 256
   E. 284

2. What is the perimeter of the trapezoid shown below?

   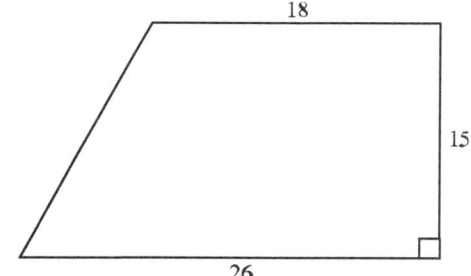

   F. 76
   G. 79
   H. 84
   J. 86
   K. 92

Trapezoids
Geometry Problem Set 28

3. An isoceles trapezoid is shown below, with lengths given in centimeters. What is the area, in square centimeters, of the trapezoid?

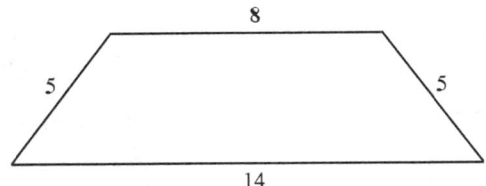

- A. 11
- B. 44
- C. $11\sqrt{61}$
- D. 88
- E. $22\sqrt{61}$

4. In the figure below, what is the value of x in trapezoid $ABCD$?

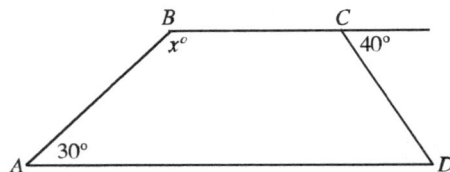

- F. 120
- G. 130
- H. 150
- J. 160
- K. 170

Trapezoids
Geometry Problem Set 28

5. In the figure below, the equilateral triangle ECD is inscribed within trapezoid $ABCD$. What is the measure of $\angle BCE$?

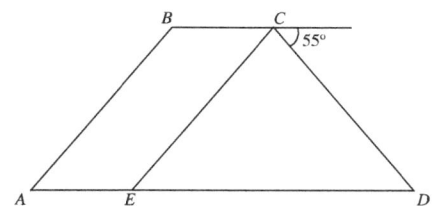

- A. $35°$
- B. $45°$
- C. $65°$
- D. $100°$
- E. $125°$

6. What is the area of trapezoid $ABCD$, shown below?

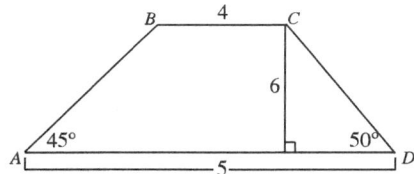

- F. 24
- G. 27
- H. 30
- J. 54
- K. 60

Trapezoids
Geometry Problem Set 28

7. What is the perimeter of the trapezoid shown below?

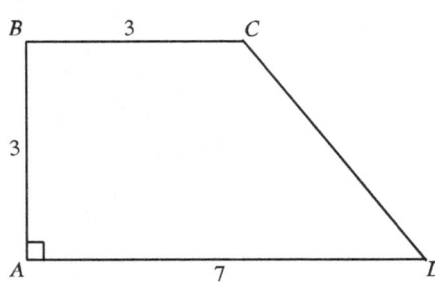

- A. 13
- B. 15
- C. 18
- D. 19
- E. 30

8. In the figure below, the length of $\overline{NO}$ is 12 and the length of each side of quadrilateral $LMNP$ is 13. What is the area of trapezoid $LMNO$?

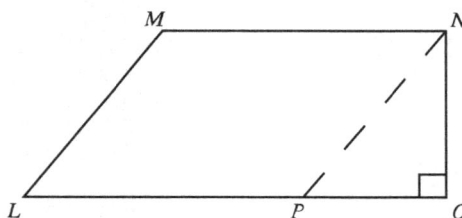

- F. 156
- G. 186
- H. 192
- J. 204
- K. 216

DO YOUR FIGURING HERE

Trapezoids
Geometry Problem Set 28

9. If the trapezoid shown below is isosceles, then what is the area of the trapezoid?

- A. $8\sqrt{2}$
- B. $8\sqrt{3}$
- C. $9\sqrt{2}$
- D. $9\sqrt{3}$
- E. 16

10. In the trapezoid shown below, $\overline{AB} = \overline{CD}$. $\overline{BC} = 6, \overline{CE} = 3x$, and $\overline{DE} = x$. The area is 48. What is the value of x?

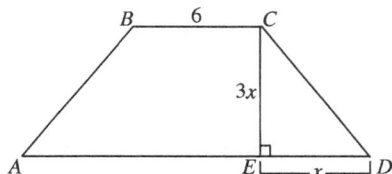

- F. 1
- G. 2
- H. 4
- J. 6
- K. 8

DO YOUR FIGURING HERE

Trapezoids
Geometry Problem Set 28

11. In the trapezoid shown below, $\overline{AB} = \overline{CD} = 5$. $\overline{BC} = 3$ and $\overline{AD} = 9$. What is the area of the trapezoid $ABCD$?

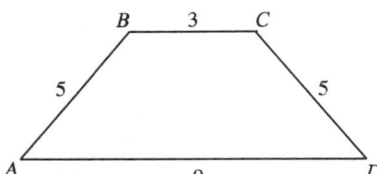

A. 18
B. 24
C. 27
D. 30
E. 48

12. The trapezoid shown below is isoceles. $\overline{AD} = (x+10)$, $\overline{BC} = x$, and $\overline{CE} = x+2$. The area of trapezoid $ABCD$ is 180. What is a possible value of x?

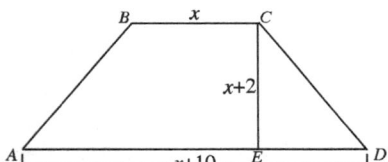

F. 4
G. 5
H. 10
J. 12
K. 20

Trapezoids
Geometry Problem Set 28

13. In the trapezoid shown below, $\angle BAD = 60°$ and $\angle ACD = 30°$. $\overline{AE} = 3$. What is the area of trapezoid $ABCD$?

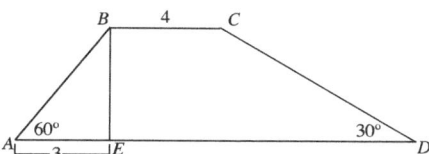

- **A.** 30
- **B.** $24\sqrt{3}$
- **C.** $30\sqrt{3}$
- **D.** 54
- **E.** 60

14. In the trapezoid shown below, $\angle BAD = 60°$ and $\angle ACD = 30°$. $\overline{AE} = 3$. What is the perimeter of trapezoid $ABCD$?

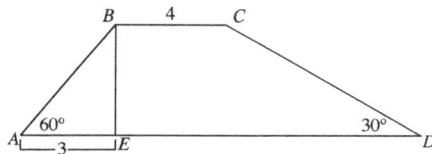

- **F.** $12 + \sqrt{3}$
- **G.** $13 + 3\sqrt{3}$
- **H.** 25
- **J.** 32
- **K.** $26 + 6\sqrt{3}$

DO YOUR FIGURING HERE

Trapezoids
Geometry Problem Set 28

Answer Key

#	Answer	Frequency	Difficulty
1	B	popular	2
2	F	popular	2
3	B	popular	2
4	H	popular	1
5	C	popular	1
6	G	popular	1
7	C	popular	2
8	G	popular	3
9	D	popular	3
10	G	popular	3
11	B	popular	3
12	H	popular	3
13	C	popular	4
14	K	popular	4

Circles I
Quick Drill

1. What is the area of a circle with a diameter of 6?

2. What is the circumference of a circle with a radius of 17?

3. What is the area of a circle with a radius of a?

4. What is the circumference of a circle with a diameter of q?

5. What is the radius of a circle with an area of $3c^2\pi$?

6. What is the radius of a circle with a circumference of 8π?

7. What is the diameter of a circle with an area of $22b^3\pi$?

8. What is the area of a circle with a circumference of 7π?

9. What is the circumference of a circle with a radius of $\dfrac{3a}{2}$?

10. What is the area of a circle with a diameter of 12?

Circles I
Quick Drill

Answer Key

#	Answer
1	9π
2	34π or
3	$a^2\pi$
4	$q\pi$
5	$c\sqrt{3}$
6	4
7	$2b\sqrt{22b}$
8	$\frac{49}{4}\pi$ or 38.5
9	$3a\pi$
10	36π

Circles
Geometry Problem Set 29

1. What is the circumference of circle O in terms of π?

   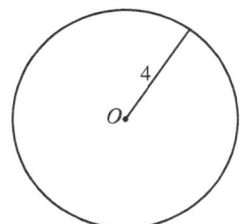

   A. 4π
   B. 8π
   C. 16π
   D. 36π
   E. 42π

2. What is the area of circle C in terms of π?

   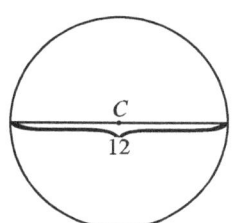

   F. 6π
   G. 12π
   H. 36π
   J. 96π
   K. 144π

3. What is the diameter of a circle with an area of 16π?

   A. 2
   B. 4
   C. 6
   D. 8
   E. 16

Circles
Geometry Problem Set 29

4. What is the radius of a circle with a circumference of 64π?

   F. 8
   G. 16
   H. 32
   J. 48
   K. 64

5. In the figure below, the circle with diameter $\overline{XY}$ is inscribed in a square with sides of length 8. What is the length of $\overline{XY}$?

   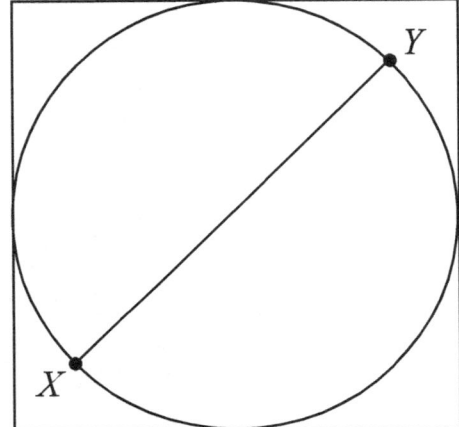

   A. 6
   B. 7
   C. 8
   D. $8\sqrt{2}$
   E. $8\sqrt{3}$

6. The circumference of circle A is 4 times the circumference of circle B. What is the ratio of the radius of circle A to the radius of circle B?

   F. 4 : 1
   G. 8 : 1
   H. 16 : 1
   J. 4 : 3
   K. 1 : 3

Circles
Geometry Problem Set 29

7. What is the ratio of the radius, r, of a circle to the circumference of the circle?

   A. $1\dfrac{\pi}{2}$

   B. $1 : \pi$

   C. $1 : 2\pi$

   D. $\pi : 1$

   E. $2\pi : 1$

8. In the figure below, A is the center of the circle and segment $\overline{XZ}$ is tangent to the circle at point Y. If the measure of $\angle XYA$ is $b°$, how many possible values are there for b?

   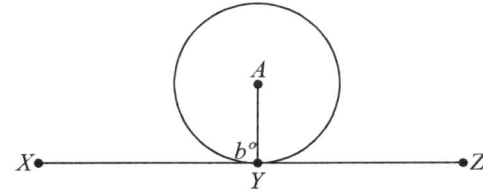

   F. 1

   G. 2

   H. 4

   J. 6

   K. More than 6

Circles
Geometry Problem Set 29

9. In the figure below, the circles are tangent as shown and the center of circle A is also the center of the largest circle. If the circumference of circle A is 2π and both circle B and circle C have circumferences of 10π, what is the area of the largest circle?

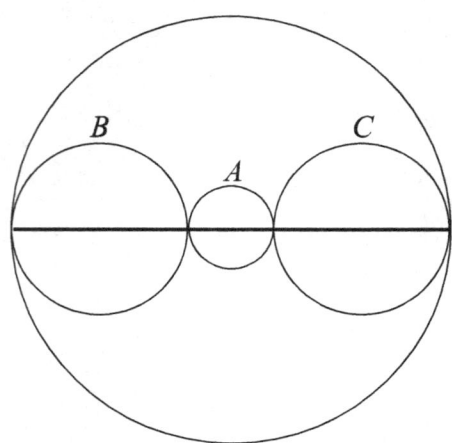

A. 36π
B. 81π
C. 100π
D. 121π
E. 144π

10. What is the area of a circle whose circumference is π?

F. $\dfrac{\pi}{4}$
G. $\dfrac{\pi}{2}$
H. $\dfrac{\pi}{8}$
J. π
K. 2π

11. A wheel made 4,000 revolutions while traveling $52,000\pi$ centimeters in a straight line along a surface. What is the radius, in centimeters, of the wheel?

Circles
Geometry Problem Set 29

12. In the figure below, A, B, and C lie on a line. B is the center of the smaller circle and C is the center of the larger circle. If the radius of the larger circle is 12, what is the area of the smaller circle?

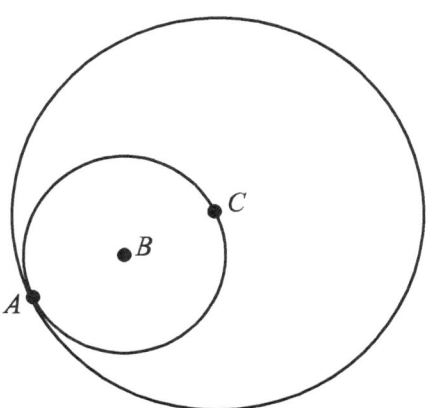

F. 9π
G. 16π
H. 36π
J. 64π
K. 144π

Circles
Geometry Problem Set 29

13. In the figure below, a square with sides 12 inches long is divided into 9 smaller squares. What is the area of the circle (not shown) that passes through points w, x, y, and z, which are the centers of the four small corner squares?

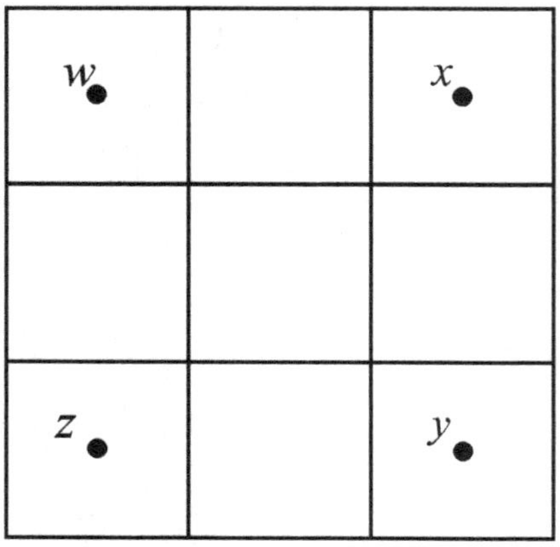

    A. 16π
    B. 32π
    C. 36π
    D. 48π
    E. 64π

14. A man rides on a unicycle at a constant speed for 200 feet. If the radius of the wheel of his unicycle is 1.5 feet, to the nearest whole number, what is the number of revolutions the wheel of the unicycle makes during his ride?

    F. 15
    G. 21
    H. 22
    J. 42
    K. 43

Circles
Geometry Problem Set 29

15. The centers of the three circles below lie on segment $\overline{AB}$ (not shown), and the three circles are mutually tangent at point B. The center of the largest circle is Point A, and the center of the middle circle lies on the smallest circle. If the radius of the smallest circle is 4, what is the area of the largest circle?

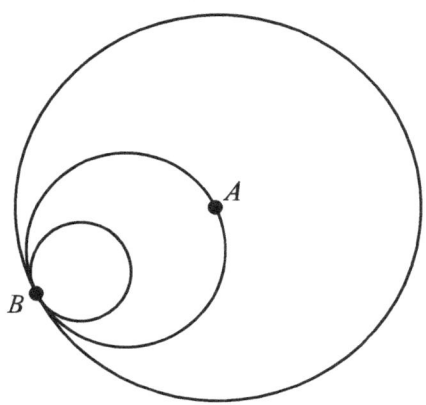

A. 16π
B. 25π
C. 64π
D. 225π
E. 256π

16. In the xy-coordinate plane, the center of a circle has coordinates $(-3, 8)$, and the circle touches the y-axis at one point only. What is the radius of the circle?

Circles
Geometry Problem Set 29

Answer Key

#	Answer	Frequency	Difficulty
1	B	popular	1
2	H	popular	1
3	D	popular	1
4	H	popular	1
5	C	popular	1
6	F	popular	1
7	C	popular	2
8	F	popular	1
9	D	popular	3
10	F	popular	2
11	6.5 cm	popular	3
12	H	popular	3
13	B	popular	4
14	G	popular	3
15	E	popular	3
16	3	average	3

Circles II
Quick Drill

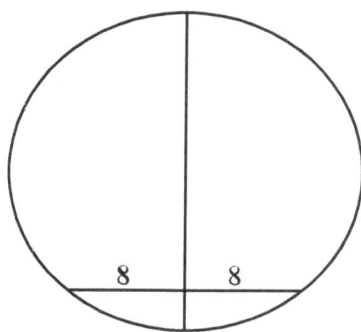

1. The distance from the center of the circle to the chord is 6. Find the area of the circle.

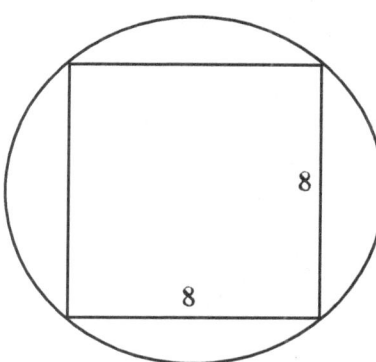

3. Find the area of the square and the area of the circle.

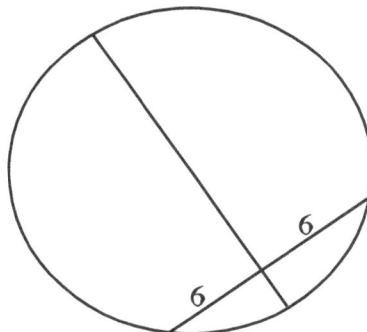

2. The distance from the center of the circle to the chord is 6. Find the area of the circle.

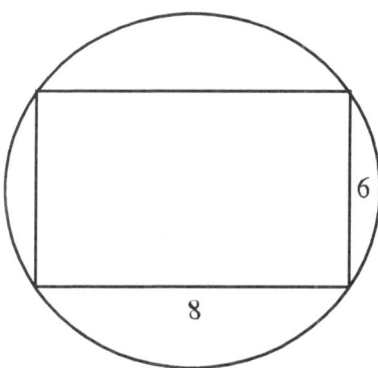

4. Find the area of the rectangle and the area of the circle.

Circles II
Quick Drill

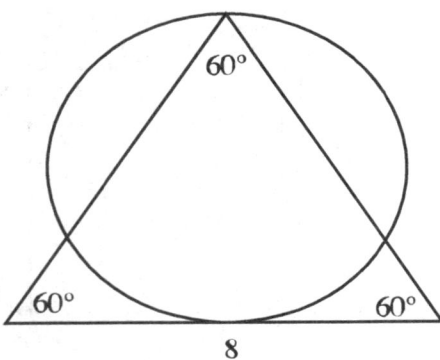

5. Find the area of the triangle and the circumference of the circle.

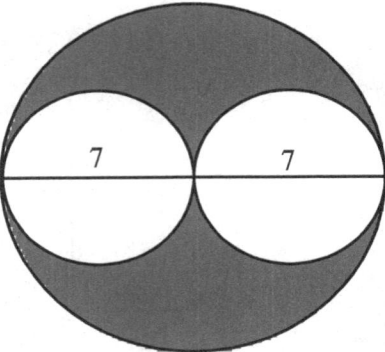

7. Find the area of the shaded region.

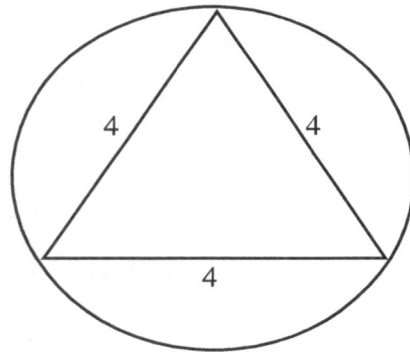

6. Find the area of the triangle and the circumference of the circle.

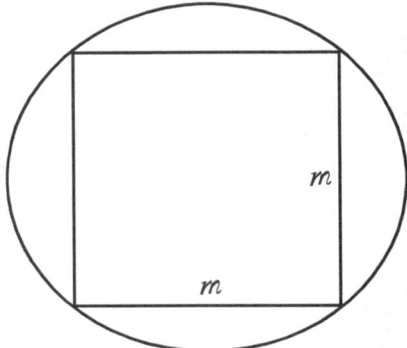

8. Find the area of the square and the area of the circle in terms of m.

Circles II
Quick Drill

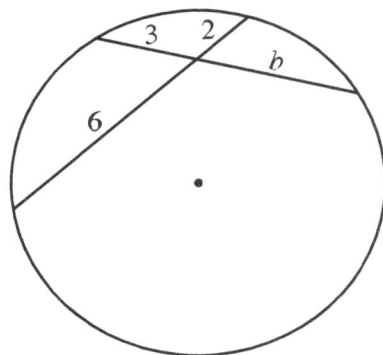

9. What is the value of b in the figure below?

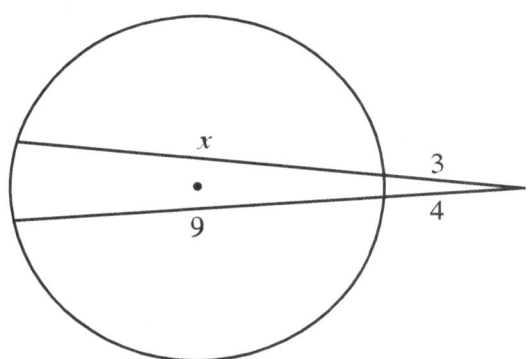

10. What is the value of x in the figure below?

Circles II
Quick Drill

Answer Key

#	Answer
1	100π
2	72π
3	Square Area: 64 and Circle Area: 32π
4	Rectangle Area: 48 and Circle Area: 25π
5	Triangle Area: $16\sqrt{3}$ and Circle Circumference: $4\sqrt{3}\pi$
6	Triangle Area: $4\sqrt{3}$ and Circle Circumference: $\frac{4\sqrt{3}}{3}\pi$
7	$\frac{49\pi}{2}$
8	Square: m^2 and Circle: $\frac{m^2}{2}\pi$
9	$b = 4$
10	$x = 9$

Inscribed Triangles
Geometry Problem Set 30

1. In the figure below, if P is the center of the circle, what is the value of a?

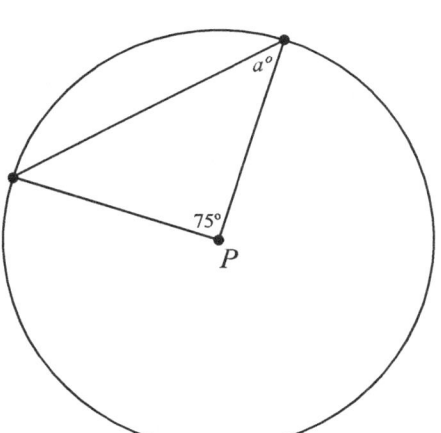

2. In the figure below, the circle has center O. If the measure of $\angle SOP = 84°$, what is the value of w?

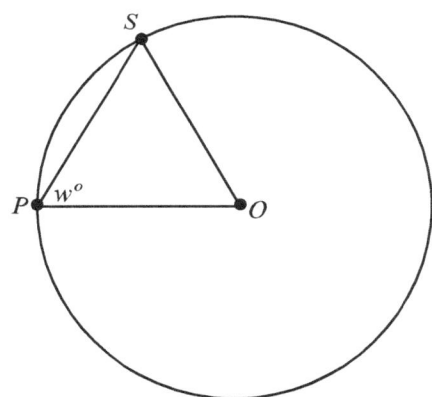

F. 48
G. 53
H. 58
J. 64
K. 96

Inscribed Triangles
Geometry Problem Set 30

3. B is the center of the circle shown below, and the circle has a circumference of 12π. What is the area of the $\triangle ABC$? (Circumference = πD)

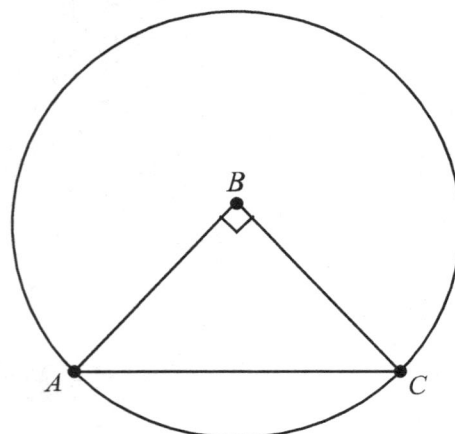

A. 12
B. 18
C. 24
D. 36
E. 48

4. Y is the center of the circle shown below, and the length of $\overline{XZ}$ is $\sqrt{32}$. What is the circumference of the circle?

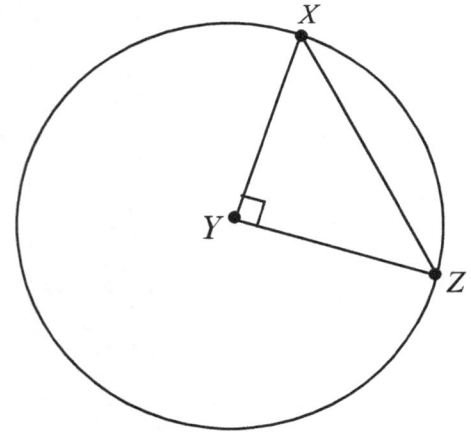

F. 8π
G. 12π
H. 16π
J. 24π
K. 32π

Inscribed Triangles
Geometry Problem Set 30

5. In the figure below, if the circle with center O has area $a^2\pi$, what is the area of $\triangle ABO$?

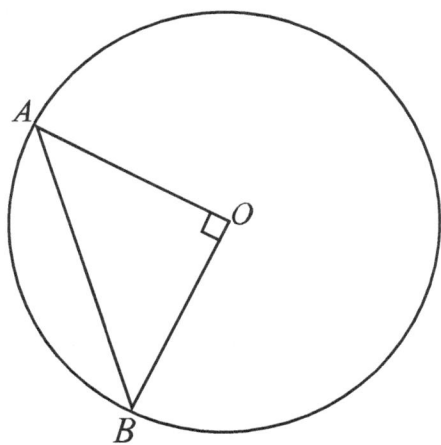

- A. $\dfrac{a}{2}$
- B. $\dfrac{a^2}{2}$
- C. $2a$
- D. a^2
- E. $2a^2$

Inscribed Triangles
Geometry Problem Set 30

6. The circle shown below with Center O has an inscribed angle of $42°$. What is the degree measure of arc ABC?

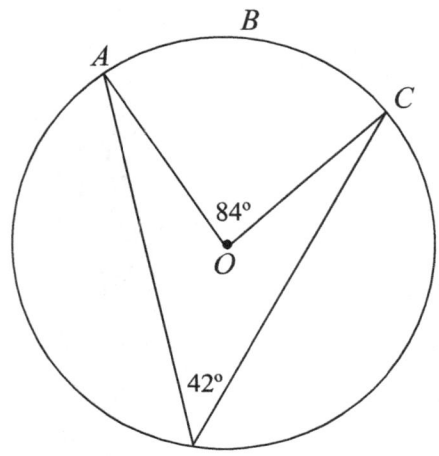

F. $42°$
G. $63°$
H. $84°$
J. $105°$
K. $126°$

DO YOUR FIGURING HERE

7. The triangle shown below is inscribed in a circle (not shown) so that points A, B, and C all lie on the circumference. If $\overline{AB}$ is a diameter, what is the length of $\overline{CB}$?

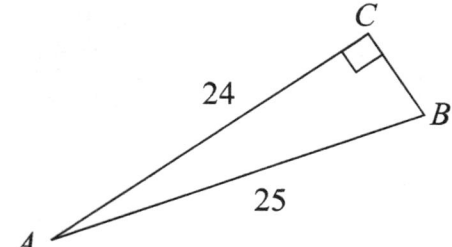

Inscribed Triangles
Geometry Problem Set 30

Answer Key

#	Answer	Frequency	Difficulty
1	52.5	popular	2
2	F	popular	2
3	B	popular	2
4	F	rare	3
5	B	average	2
6	H	average	3
7	7	average	2

Circles III
Quick Drill

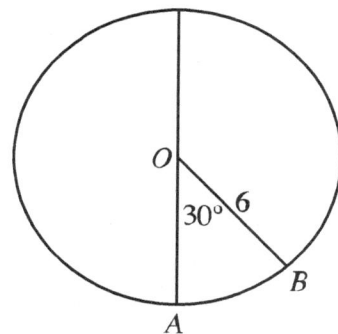

1. Find the length of arc AB and the area of sector AOB.

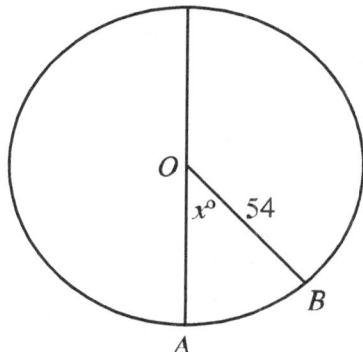

3. Find x when the length of arc AB is 12π.

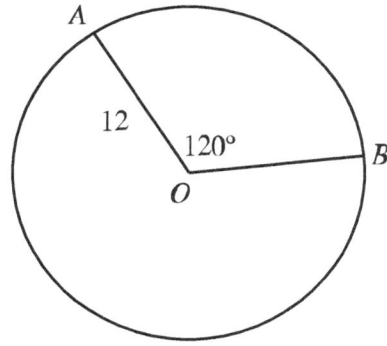

2. Find the length of arc AB and the area of sector AOB.

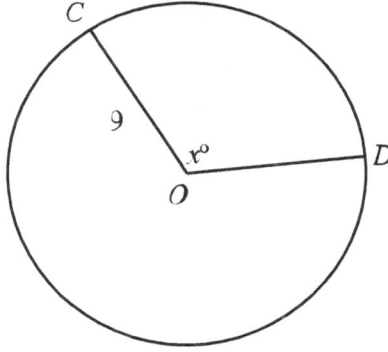

4. Find x when the area of sector COD is 27π.

Circles III
Quick Drill

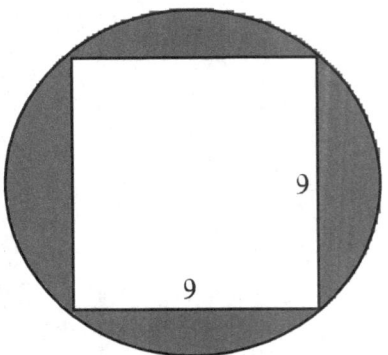

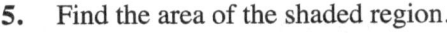

5. Find the area of the shaded region.

6. If a sector has a central angle of 20° in a circle with a radius of 6, then what is the arc length and sector area?

7. If a circle has an arc with a length of 10π and a central angle of 36° then what is the radius of the circle?

8. If a circle has a sector with an area of 27π and a central angle of 270° then what is the circumference of the circle?

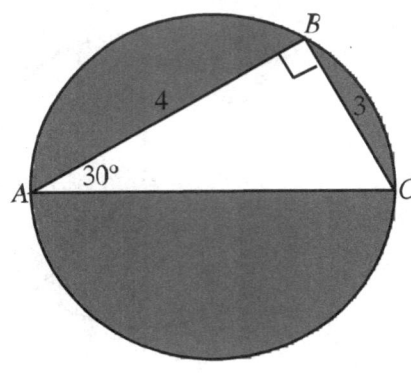

9. What is the length of arc BC?

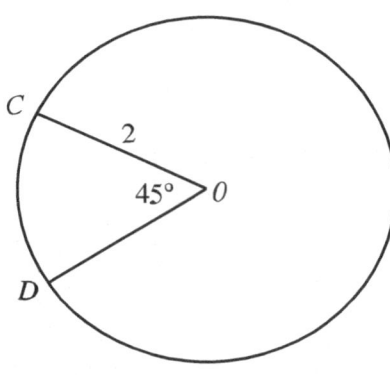

10. Find the length of arc CD and the area of sector COD.

Circles III
Quick Drill

Answer Key

#	Answer	Frequency	Difficulty
1	Arc Length: π and Sector Area: 3π	popular	1
2	Arc Length: 8π and Sector Area: 48π	popular	1
3	$x = 40°$	popular	1
4	$x = 120°$	popular	1
5	$\dfrac{81\pi}{2} - 81$	popular	1
6	Arc Length: $\dfrac{2\pi}{3}$ and Sector Area: 2π	popular	1
7	$r = 50$	popular	1
8	12π	popular	1
9	$\dfrac{5}{12}\pi$	popular	1
10	Arc Length: $\dfrac{\pi}{2}$ and Sector Area: $\dfrac{\pi}{2}$	popular	1

Circle Sectors
Geometry Problem Set 31

1. If the shaded region of the circle with center O is $\frac{7}{45}$ of the total area of the circle, what is the measure of $\angle NOP$?

   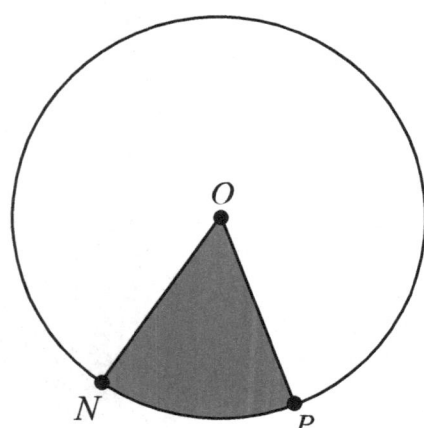

   A. $48°$
   B. $52°$
   C. $56°$
   D. $60°$
   E. $64°$

2. A pizza is cut into 12 equal slices. If one person eats five of these slices, what is the total degree measure of the arc made by the crusts of these five slices?

   F. $130°$
   G. $145°$
   H. $150°$
   J. $180°$
   K. $210°$

Circle Sectors
Geometry Problem Set 31

3. In the figure below, the length of the darkened arc PQ is $\frac{1}{8}$ of the circumference of the circle with center O. If the radius of the circle is 12, what is the length of the arc?

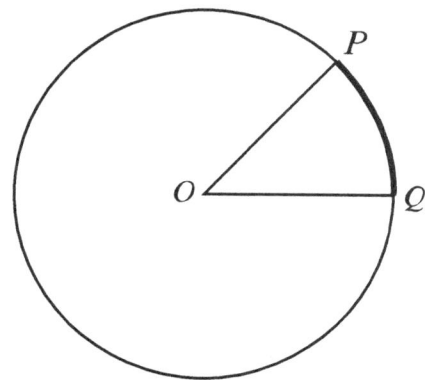

A. $\frac{\pi}{2}$

B. $\frac{3\pi}{2}$

C. 2π

D. 3π

E. 6π

Circle Sectors
Geometry Problem Set 31

4. The circle below has an area of 16π and is divided into 6 congruent regions. What is the perimeter of one of these regions?

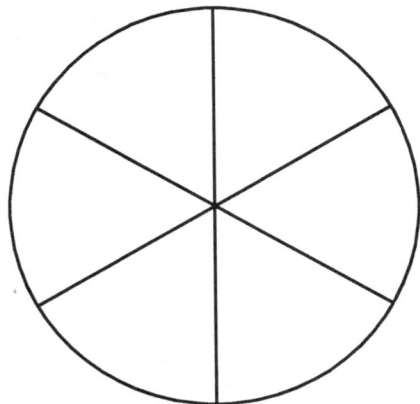

F. $8 + 4\pi$

G. $8 + 16\pi$

H. $8 - 16\pi$

J. $8 + \dfrac{4}{3}\pi$

K. $8 + \dfrac{2}{3}\pi$

5. The minute hand of a circular clock on the wall makes one complete revolution each hour. Through how many degrees does the minute hand turn in the 12 minutes from 10:00 to 10:12?

A. 72°

B. 76°

C. 82°

D. 88°

E. 108°

Circle Sectors
Geometry Problem Set 31

6. The circle shown below is centered at point O. If the measure of angle MON is $120°$, what is the total area of the shaded region?

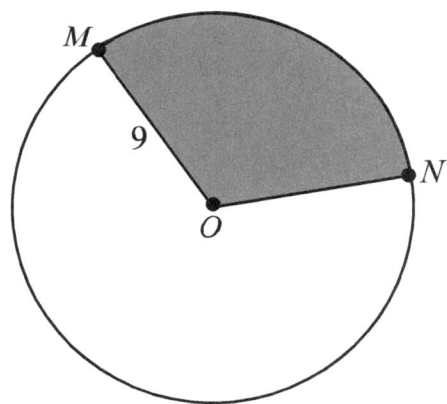

F. 9π

G. 12π

H. 18π

J. 27π

K. 54π

Circle Sectors
Geometry Problem Set 31

Answer Key

#	Answer	Frequency	Difficulty
1	C	average	1
2	H	average	2
3	D	average	2
4	J	average	3
5	A	average	2
6	J	average	2

Shaded Region
Geometry Problem Set 32

1. In the figure below, D is the midpoint of $\overline{AE}$ and C is the midpoint of $\overline{BE}$. If the length of $\overline{AD}$ in the figure above is 5, what is the total area of the shaded quadrilateral $ABCD$?

   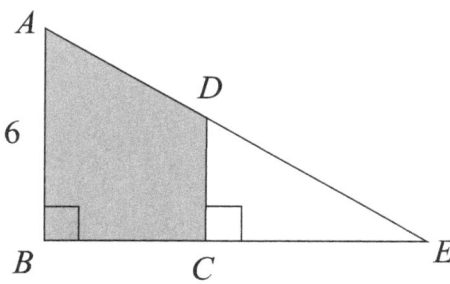

   A. 12
   B. 18
   C. 20
   D. 21
   E. 24

2. In the rectangle below, the radius of each quarter circle is 4. What is the area of the shaded region?

   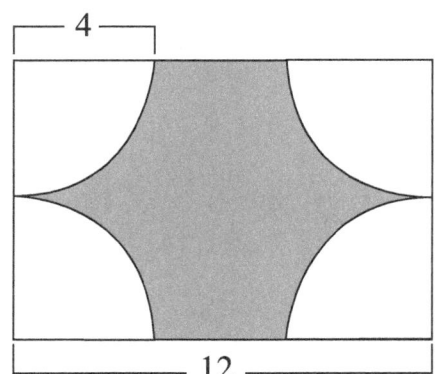

   F. $96 - 16\pi$
   G. $48 - 16\pi$
   H. 16π
   J. 96
   K. 48

Shaded Region
Geometry Problem Set 32

3. The figure below shows a trapezoid divided into three triangles. What is the area of the shaded triangle?

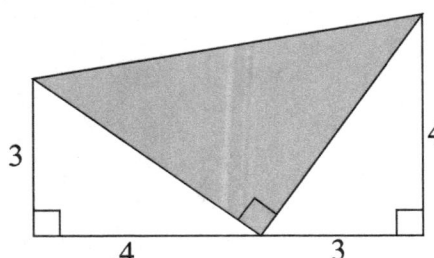

   A. 10.5
   B. 12.5
   C. 14.5
   D. 16.5
   E. 18.5

4. In the rectangle below, the sum of the areas of the shaded regions is 2. What is the area of the unshaded region?

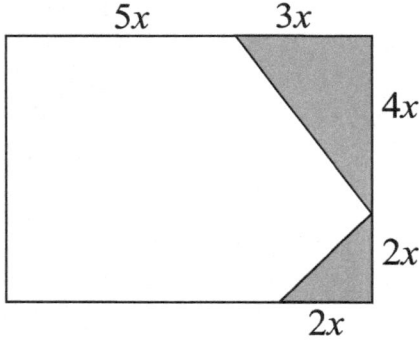

   F. 2
   G. 4
   H. 6
   J. 9
   K. 10

Shaded Region
Geometry Problem Set 32

5. The circumference of the circle shown below is 18π, and the circle is tangent to the square, as depicted. Which of the following gives the area of the shaded region?

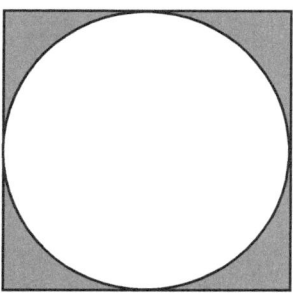

- A. $81 - 18\pi$
- B. 81
- C. $324 - 18\pi$
- D. $324 - 81\pi$
- E. 324

6. In the figure below, the 5 circles have the same center and their radii are 1, 2, 3, 4, and 5, respectively. What is the ratio of the area of the smallest unshaded ring to the area of the largest shaded ring?

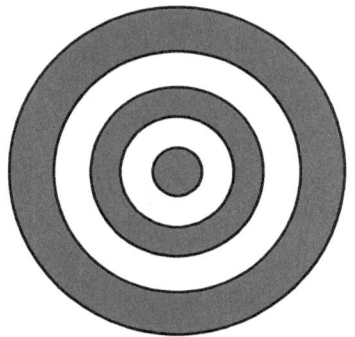

- F. $1:3$
- G. $2:3$
- H. $4:9$
- J. $5:9$
- K. $4:25$

Shaded Region
Geometry Problem Set 32

7. In the figure below, a circle with center O is inscribed in a square. What is the area of the shaded portion of the circle?

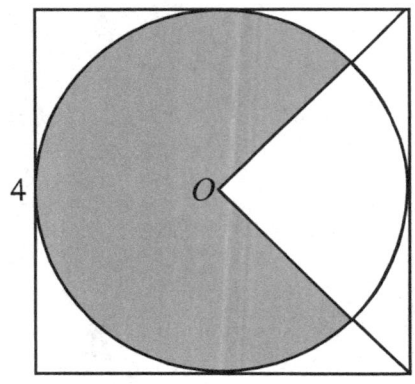

DO YOUR FIGURING HERE

A. π
B. 3π
C. 4π
D. 8π
E. 12π

8. In the rectangle $PQRS$ below, the area of the shaded region is given by $A = \dfrac{\pi lw}{4}$. If the area of the shaded region is 9π, what is the total area, to the nearest whole number, of the unshaded regions of rectangle $PQRS$?

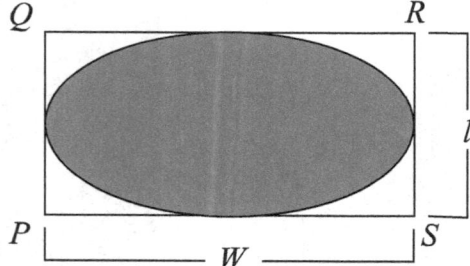

F. 8
G. 10
H. 12
J. 24
K. 36

Shaded Region
Geometry Problem Set 32

9. In the figure below, the coordinates of K are $(10, 2b)$ and the coordinates of J are $(10-b, 0)$. When a point in square $LMNO$ is chosen at random, the probability that the point will be in the shaded area is $\frac{1}{5}$. What is the value of b?

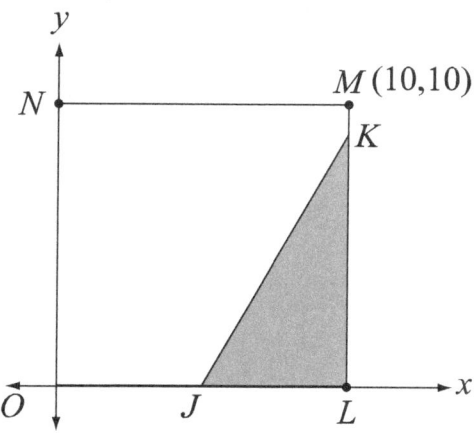

A. $\sqrt{10}$

B. $5\sqrt{2}$

C. $2\sqrt{5}$

D. $2\sqrt{10}$

E. 5

10. In the rectangle $PQRS$ below, W is the midpoint of $\overline{PS}$, Z is the midpoint of PW, Y is the midpoint of $\overline{QR}$ and X is the midpoint of $\overline{RS}$. What fraction of the area of the rectangle is shaded?

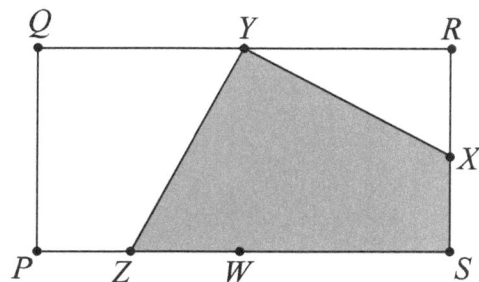

Shaded Region
Geometry Problem Set 32

Answer Key

#	Answer	Frequency	Difficulty
1	B	popular	1
2	F	popular	3
3	B	popular	2
4	K	popular	3
5	D	popular	2
6	F	popular	3
7	B	popular	2
8	F	popular	3
9	C	popular	3
10	$\frac{1}{2}$	popular	2

Plane Geometry Mixed Problem Set
Geometry Problem Set 33

1. What is the value of $x - y$?

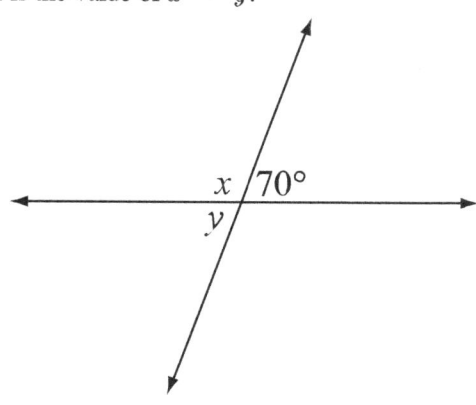

   A. 40°
   B. 60°
   C. 70°
   D. 80°
   E. 90°

2. According to the figure below, what is the value of a?

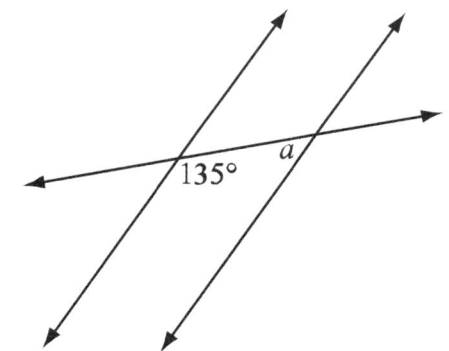

   F. 30°
   G. 45°
   H. 60°
   J. 90°
   K. 135°

Plane Geometry Mixed Problem Set
Geometry Problem Set 33

3. In the figure below, what is the value of z?

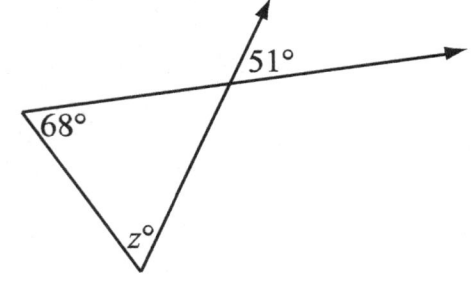

DO YOUR FIGURING HERE

A. 51
B. 56
C. 61
D. 68
E. 72

4. In the figure below, what is the value of w?

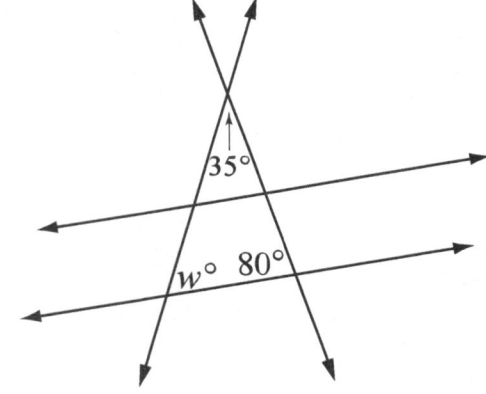

F. 65
G. 75
H. 80
J. 85
K. 90

Plane Geometry Mixed Problem Set
Geometry Problem Set 33

5. In the figure below, what is the value of x?

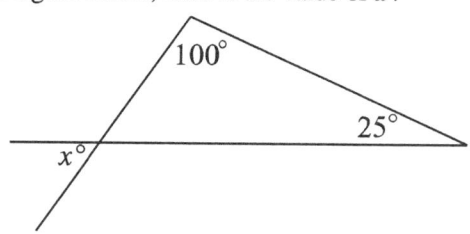

A. 40
B. 45
C. 50
D. 55
E. 60

6. Lines l, m and n intersect at a point shown below. If $a = 65$, what is the value of b?

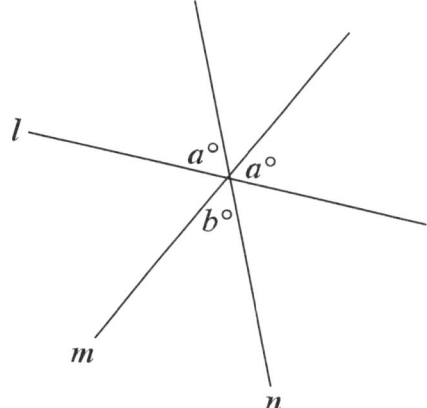

F. 25
G. 50
H. 55
J. 60
K. 100

Plane Geometry Mixed Problem Set
Geometry Problem Set 33

7. If $\overline{AB} \parallel \overline{DC}$ in the figure above, what is the value of $\dfrac{x+y}{2}$?

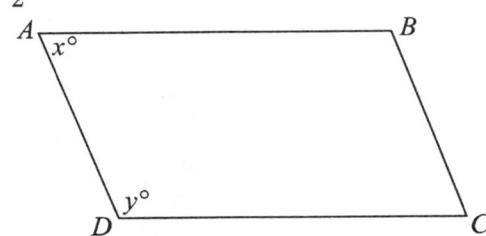

A. 45
B. 60
C. 90
D. 120
E. 145

8. If $\overline{RS} \parallel \overline{PQ}$ in the figure below, what is the value of $\dfrac{a+b}{2}$?

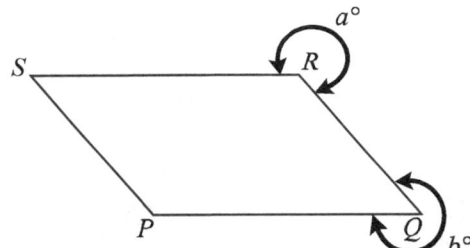

F. 45
G. 60
H. 90
J. 120
K. 270

Plane Geometry Mixed Problem Set
Geometry Problem Set 33

9. In the figure below, the circle has center O. If the measure of arc PAS is $74°$, what is the value of w?

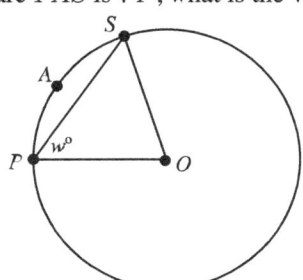

   A. 53
   B. 57
   C. 65
   D. 74
   E. 105

10. In the figure below, square $NEAT$ is inscribed in a circle. What is the degree measure of arc NE?

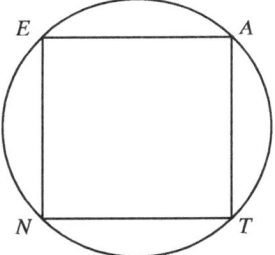

   F. 45°
   G. 60°
   H. 90°
   J. 120°
   K. 180°

11. In the xy-coordinate plane, what is the area of the square with opposite vertices at $(-4,-4)$ and $(4,4)$?

   A. 16
   B. 36
   C. 49
   D. 64
   E. 81

Plane Geometry Mixed Problem Set
Geometry Problem Set 33

12. In the figure below, $\overline{JK}$ is a diagonal of a square (not shown). Which of the following are the coordinates of one of the end points of the other diagonal of the square?

    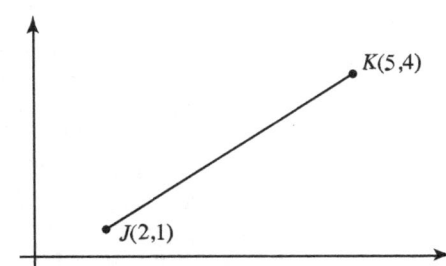

    F. (2, 5)

    G. (1, 5)

    H. (2, 4)

    J. (4, 2)

    K. (2, 5)

13. In $\triangle ABC$, the length of side $\overline{AB}$ is 10 and the length of side $\overline{BC}$ is 18. What is the smallest possible integer length of side $\overline{AC}$?

    A. 8

    B. 9

    C. 14

    D. 27

    E. 28

14. The length of the longest side of a triangle is 7, and the remaining two sides have integer lengths. If the lengths of all three sides are different integers, what is one value that cannot be the perimeter of the triangle?

    F. 14

    G. 15

    H. 16

    J. 17

    K. 18

Plane Geometry Mixed Problem Set
Geometry Problem Set 33

15. In the figure below, what is the area of square $ABCD$?

    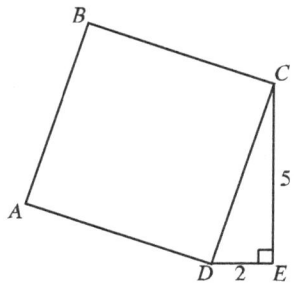

    A. 21
    B. 25
    C. 29
    D. 35
    E. 49

16. What is the greatest possible area of a triangle with one side of length 5 and another side of length 5?

    F. 10
    G. 12.5
    H. 20
    J. 25
    K. 30

Plane Geometry Mixed Problem Set
Geometry Problem Set 33

17. In the figure below, $\overline{XZ}$ passes through point P, and $\overline{PY}$ is perpendicular to $\overline{PW}$. What is the measure of $\angle XPW$?

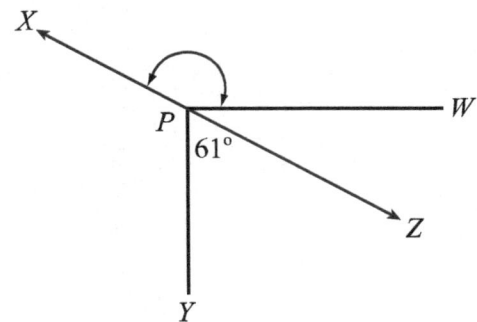

- A. 119°
- B. 131°
- C. 133°
- D. 151°
- E. 161°

18. In the figure below, line w is perpendicular to line z, and line y bisects the angle with measure $a°$. In terms of b what does c equal?

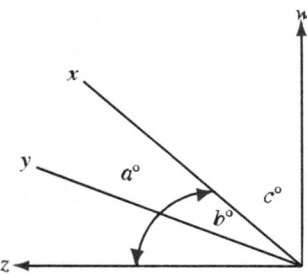

- F. $2b$
- G. $180 - 5b$
- H. $25 + b$
- J. $90 - 2b$
- K. $90 - b$

Plane Geometry Mixed Problem Set
Geometry Problem Set 33

19. Three different lines are drawn in a plane. What is the smallest number of regions into which these lines will divide the plane?

    A. 2
    B. 3
    C. 4
    D. 5
    E. 6

20. Two lines are drawn in a plane. Into how many non-overlapping regions can the plane be divided?

    F. 2 only
    G. 3 only
    H. 4 only
    J. 3 or 4
    K. 2, 3 or 4

21. If $a = 40°$ and $b = 30°$ in the figure shown below, what is the value of c?

    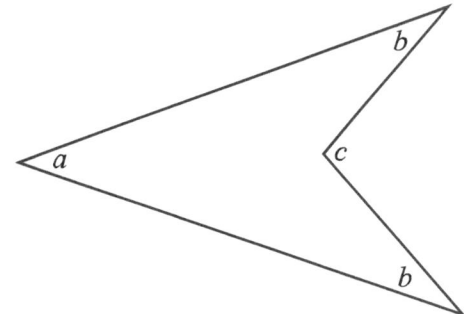

    A. 70°
    B. 80°
    C. 90°
    D. 100°
    E. 110°

Plane Geometry Mixed Problem Set
Geometry Problem Set 33

22. In the figure below, $KLNO$ is a parallelogram and $KMNP$ is a rectangle. If $\overline{LN} = 18$, $\overline{KM} = 8$ and $\overline{LM} = 3$, what is the area of rectangle $KMNP$?

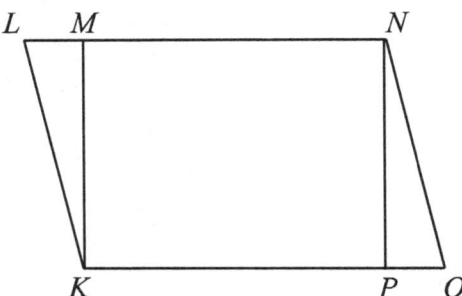

DO YOUR FIGURING HERE

F. 75
G. 90
H. 120
J. 144
K. 165

23. In the figure below, NOP is an equilateral triangle and $LMNO$ is a square with an area of 4. What is the perimeter of polygon $LMNPO$?

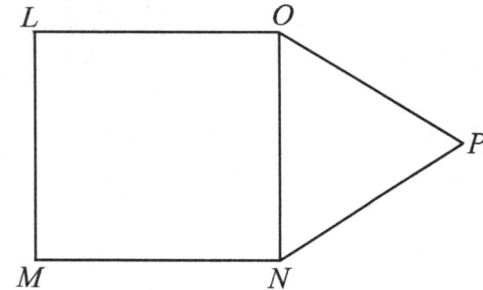

A. 8
B. 10
C. 12
D. 14
E. 16

Plane Geometry Mixed Problem Set
Geometry Problem Set 33

24. In the figure below, what is the length of segment $\overline{AB}$?

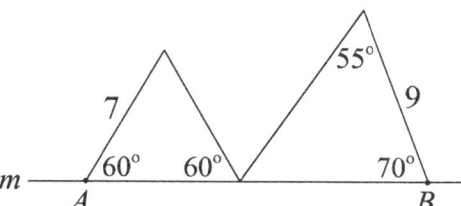

- **F.** 14
- **G.** 15
- **H.** 16
- **J.** 17
- **K.** 18

25. In the figure below, a small circle is inside a large square. What is the area, in terms of x, of the shaded region?

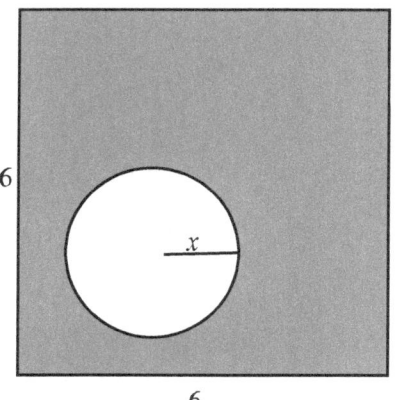

- **A.** $36 - 2\pi x$
- **B.** $2\pi x - 12$
- **C.** $36 - x^2$
- **D.** $12 - \pi x^2$
- **E.** $36 - \pi x^2$

Plane Geometry Mixed Problem Set
Geometry Problem Set 33

26. Point O is the center of both circles in the figure below. If the circumference of the small circle is 36 and the radius of the small circle is half of the radius of the large circle, what is the length of the darkened arc?

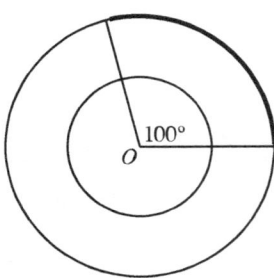

F. 10
G. 20
H. 8π
J. 36
K. 48

27. Which of the following must be true about w, x, and z in the figure shown below?

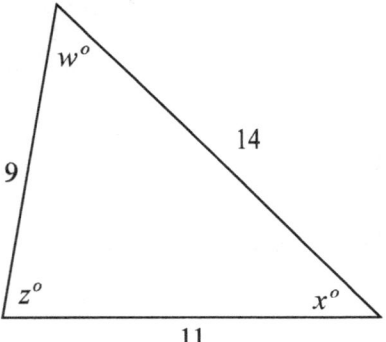

A. $x > z > w$
B. $w > z > x$
C. $z > w > x$
D. $z > x > w$
E. $w > x > z$

Plane Geometry Mixed Problem Set
Geometry Problem Set 33

28. If the information in the figure displayed below is accurate, which of the following must be true about n?

I. $0 < n < 60$
II. $l > n$
III. $n > m$

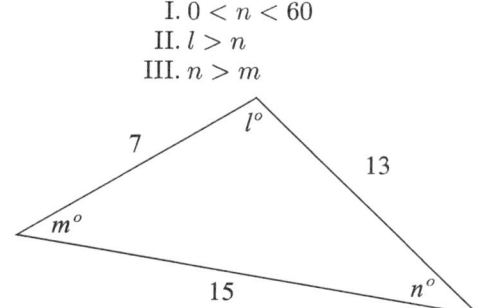

F. I only

G. II only

H. III only

J. I and II only

K. I, II, and III

29. In $\triangle ABC$, $\overline{AB} = 5$, $\overline{BC} = 12$, $\overline{AC} = 14$, and the measure of $\angle ABC$ is $b°$. Which of the following must be true about b?

A. $b = 90$

B. $b > 90$

C. $60 < b < 90$

D. $b = 60$

E. $0 < b < 60$

Plane Geometry Mixed Problem Set
Geometry Problem Set 33

30. In the figure below, line m passes through two vertices of a square and then through one vertex of a rectangle. Which of the following represents the sum of the areas of the shaded regions?

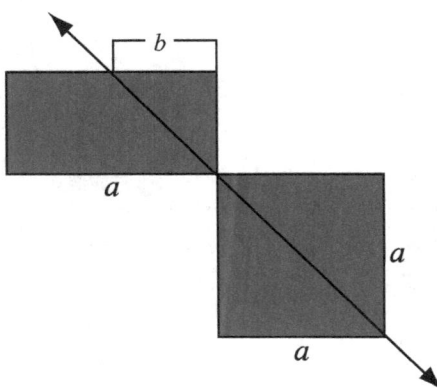

- **F.** $a^2 + \sqrt{ab}$
- **G.** $a^2 + ab$
- **H.** $a^2 + 2ab$
- **J.** $2a^2 + \sqrt{ab}$
- **K.** $2a^2 + ab$

31. The figure below shows three squares with sides of length 4, 6, and x. If A, B, and C lie on line l, what is the value of x?

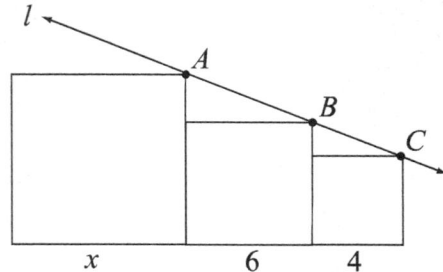

Plane Geometry Mixed Problem Set
Geometry Problem Set 33

32. The regular hexagon FGHIJK is inscribed in a circle. What is the ratio of the length of arc FGH to the length of arc FIG?

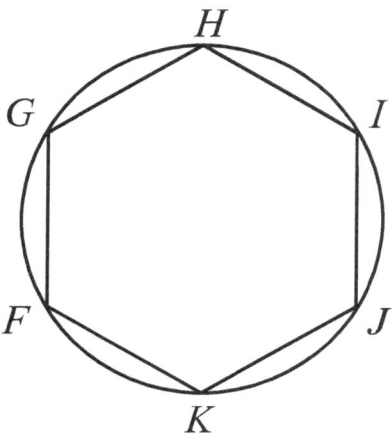

DO YOUR FIGURING HERE

F. 2 to 5
G. 1 to 3
H. 1 to 2
J. 1 to 5
K. 2 to 1

33. Pentagon PQRST in the figure below is a regular pentagon. The measure of ∠PTQ is 36°. If the measure of ∠TRQ is $x°$ and the measure of ∠TSR is $y°$ what is the value of $y-x$?

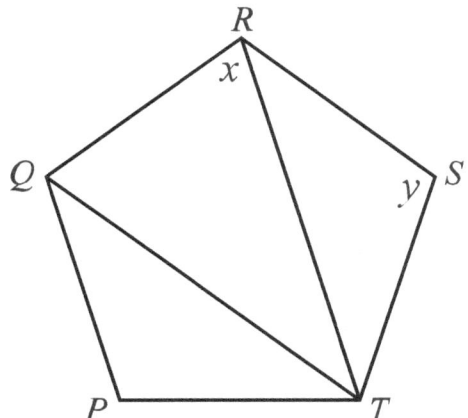

A. 36°
B. 48°
C. 56°
D. 72°
E. 108°

Plane Geometry Mixed Problem Set
Geometry Problem Set 33

34. If the information in the figure displayed below is accurate, which of the following must be true?

  I. $0 < n < 60$
  II. $p > n$
  III. $n > m$

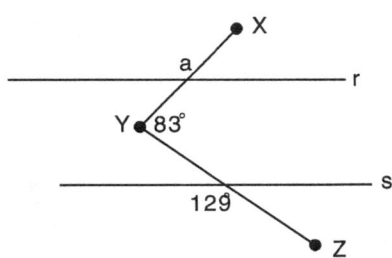

- **F.** I only
- **G.** II only
- **H.** III only
- **J.** I and II only
- **K.** I, II and III

35. In the figure below, $r \parallel s$, and segments $\overline{XY}$ and $\overline{YZ}$ intersect at a point Y. What is the value of a?

- **A.** 83
- **B.** 106
- **C.** 129
- **D.** 136
- **E.** 148

Plane Geometry Mixed Problem Set
Geometry Problem Set 33

36. If this page were folded along the dotted line in the figure below, the top half of the letter E would exactly coincide with the bottom half of E. Which of the following letters, as shown, CANNOT be folded along a horizontal line so that its top half would coincide with its bottom half?

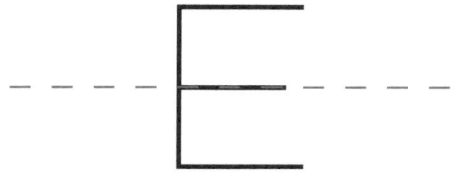

- F. C
- G. O
- H. I
- J. W
- K. H

37. In the figure below, $ABCD$ is a rectangle with $\overline{AB} = 6$ and $\overline{BC} = 8$. Points X, Y and Z are different points on a line (not shown) that is parallel to line $\overline{AD}$. Points X and Y are symmetric about line $\overline{AB}$ and points Y and Z are symmetric about line $\overline{CD}$. What is the length of $\overline{XZ}$?

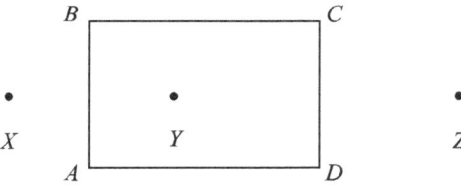

Note: Figure not drawn to scale

- A. 8
- B. 10
- C. 12
- D. 14
- E. 16

Plane Geometry Mixed Problem Set
Geometry Problem Set 33

38. If the length of a rectangle is increased by 20% and the width of the same rectangle is decreased by 20%, what is the effect on the total area of the rectangle?

  F. It is decreased by 4%

  G. It is decreased by 12%

  H. It is unchanged

  J. It is increased by 4%

  K. It is increased by 12%

39. In the figure below, line m is parallel to line l. If $b = 21$, what is the value of a?

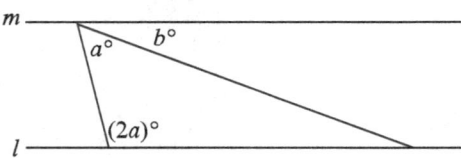

  A. 51

  B. 52

  C. 53

  D. 54

  E. 55

40. In the figure below, R is perpendicular to U and T is parallel to U. What is b in terms of a?

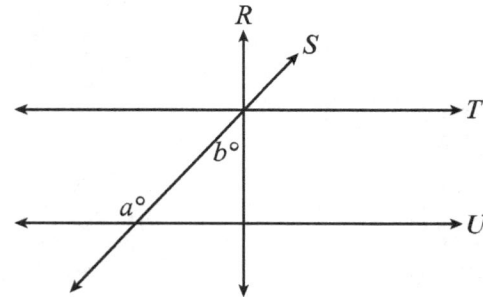

  F. $180 - 2a$

  G. $180 - a$

  H. $2a + 90$

  J. $a + 90$

  K. $a - 90$

Plane Geometry Mixed Problem Set
Geometry Problem Set 33

41. In the figure below, $PQRS$ is a square, RST is a right triangle and R is the midpoint of $\overline{TQ}$. If a point is chosen at random from polygon $SPQT$, what is the probability that the chosen point is in the shaded region?

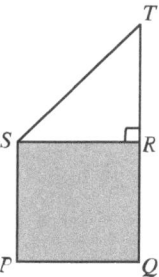

A. $\dfrac{1}{3}$

B. $\dfrac{1}{2}$

C. $\dfrac{2}{3}$

D. $\dfrac{3}{4}$

E. $\dfrac{5}{6}$

42. In the figure below, all angles are right angles and $b = 3a$. If x, a, and b are the lengths of the segments indicated, what fraction of the figure is NOT shaded?

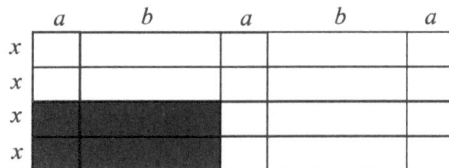

F. $\dfrac{1}{6}$

G. $\dfrac{2}{9}$

H. $\dfrac{2}{3}$

J. $\dfrac{7}{9}$

K. $\dfrac{8}{9}$

Plane Geometry Mixed Problem Set
Geometry Problem Set 33

43. In the figure below, what is the area of $\triangle QRS$?

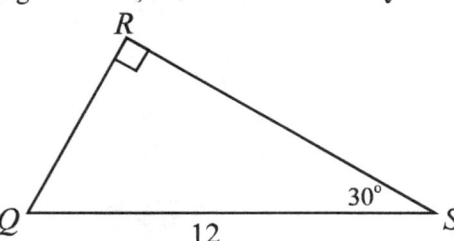

A. $3\sqrt{3}$
B. 9
C. $9\sqrt{3}$
D. $18\sqrt{3}$
E. $36\sqrt{3}$

DO YOUR FIGURING HERE

44. In the figure below, equilateral $\triangle QRS$ and equilateral $\triangle TUV$ intersect so that side $\overline{TV}$ is parallel to side $\overline{QR}$. The numbers indicate the lengths of the sides of the polygon outlined in bold. How much greater is the perimeter of $\triangle TUV$ than the perimeter of $\triangle QRS$?

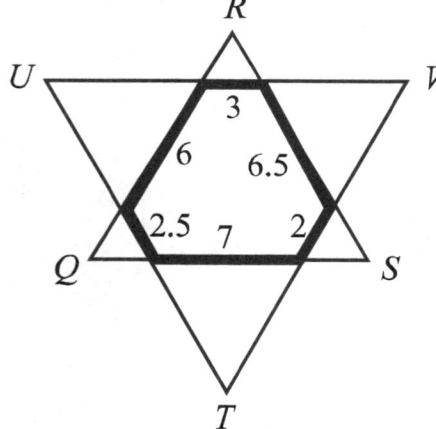

Plane Geometry Mixed Problem Set
Geometry Problem Set 33

45. In the figure below, △RST is inscribed in the circle with center O and diameter $\overline{RS}$. If $\overline{TR} = \overline{TO}$, what is the degree measure of ∠ROT?

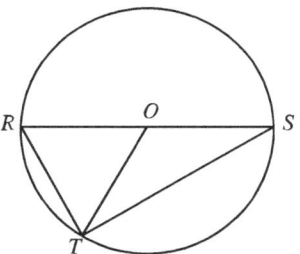

- **A.** 90°
- **B.** 60°
- **C.** 45°
- **D.** 30°
- **E.** 25°

DO YOUR FIGURING HERE

46. In the figure below, $\overline{JM}$, $\overline{NK}$, and $\overline{LO}$ intersect at A. If $b = 90$, $f = 60$, $e = 50$, $d = 50$, and $c = 45$, what is the value of a?

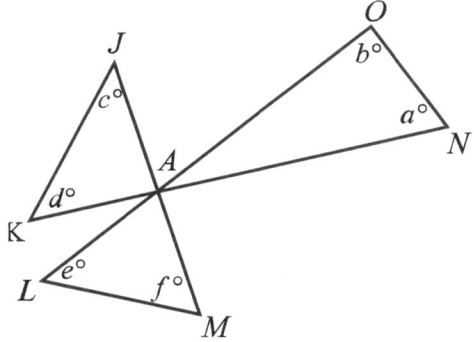

- **F.** 25
- **G.** 45
- **H.** 50
- **J.** 65
- **K.** 75

Plane Geometry Mixed Problem Set
Geometry Problem Set 33

47. The circle below has an area of 36π and is divided into 8 congruent regions. What is the perimeter of one of these regions?

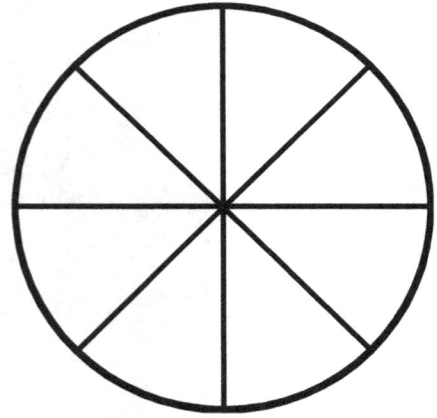

A. $12 - 64\pi$

B. $12 + \dfrac{3}{4}\pi$

C. $12 + 6\pi$

D. $12 + 64\pi$

E. $12 + \dfrac{3}{2}\pi$

DO YOUR FIGURING HERE

Plane Geometry Mixed Problem Set
Geometry Problem Set 33

48. The figure below shows an arrangement of 15 squares, each with side of length x inches. The perimeter of the figure is p inches, and the area of the figure is a inches. If $p = a$, what is the value of x?

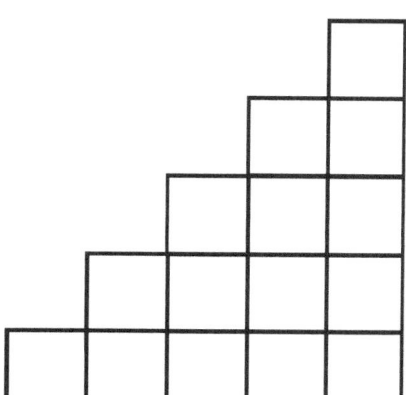

- **F.** $\dfrac{3}{4}$
- **G.** 1
- **H.** $\dfrac{4}{3}$
- **J.** $\dfrac{3}{2}$
- **K.** 2

49. In the figure above, O is the center of the circle, $OWYZ$ is a square, and $OX = 6$. What is the area of the square?

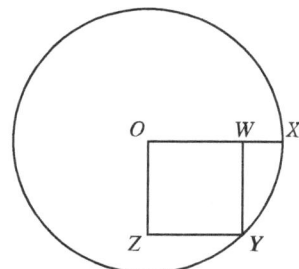

- **A.** 9
- **B.** $9\sqrt{2}$
- **C.** $9\sqrt{3}$
- **D.** 18
- **E.** 36

Plane Geometry Mixed Problem Set
Geometry Problem Set 33

50. The figure below is comprised of a trapezoid and two circles, centered at E and B. E is the midpoint of $\overline{AD}$, and $\overline{DC} \parallel \overline{AB}$. The area of circle B is 64π and the area of circle E is 16π. Angles ABC and DAB are $60°$. If the length of $\overline{DC}$ is 12, what is the perimeter of the trapezoid $ABCD$?

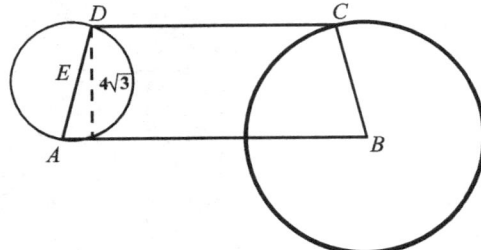

F. 32
G. $32 + 8\sqrt{3}$
H. 42
J. 48
K. 54

51. What is the area of the square shown below?

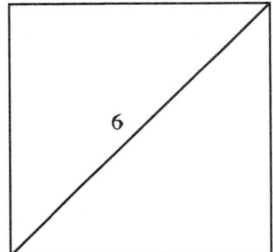

A. 18
B. 36
C. 48
D. 64
E. 128

Plane Geometry Mixed Problem Set
Geometry Problem Set 33

52. In △SPR below, line segments $\overline{SP}$ and $\overline{PR}$ are tangent to the circle at points S and Q. If the circle is centered at A, what angle must have the same measure as ∠QAR?

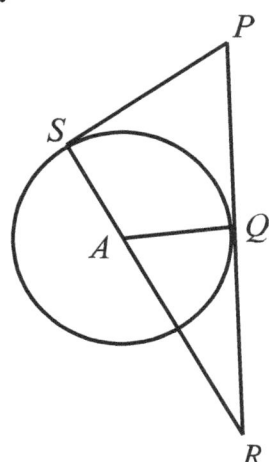

- **F.** ∠ARP
- **G.** ∠PSR
- **H.** ∠RQA
- **J.** ∠AQP
- **K.** ∠SPQ

53. The two points shown below are opposite vertices of parallelogram ABCD (not shown). If the point B (not shown) has coordinate (-2, 6), what is the remaining vertex of the parallelogram ABCD?

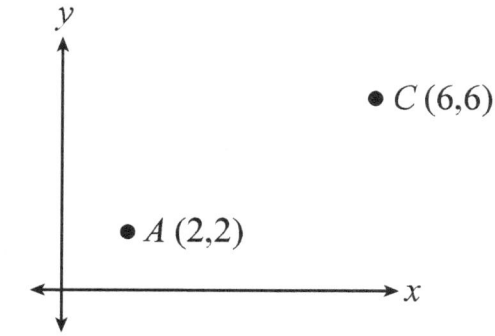

- **A.** (6, 2)
- **B.** (8, 3)
- **C.** (10, 2)
- **D.** (8, -2)
- **E.** (10, -2)

54. The pattern below is formed by four tiles measuring 3 inches by 1 inch and one square tile with side 2 inches. If a rectangular section of floor measuring 20 inches by 16 inches is to be covered with the pattern and no extra space is needed for adhesive material, how many tiles will be used?

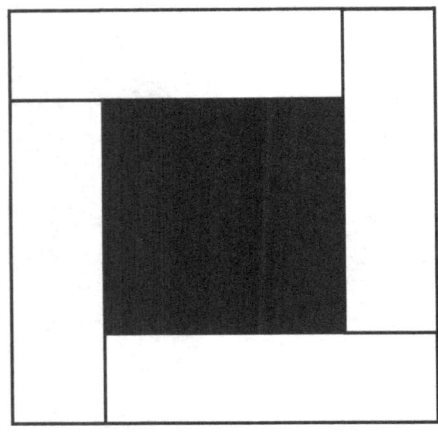

F. 20
G. 25
H. 36
J. 50
K. 100

55. Four semi-circular arcs, WX, WY, XZ and YZ divide the circle shown below into regions. The points along the circle's diameter, $\overline{WZ}$, divide the diameter into six equal parts. If $\overline{WZ} = 12$, what is the total area of the shaded regions?

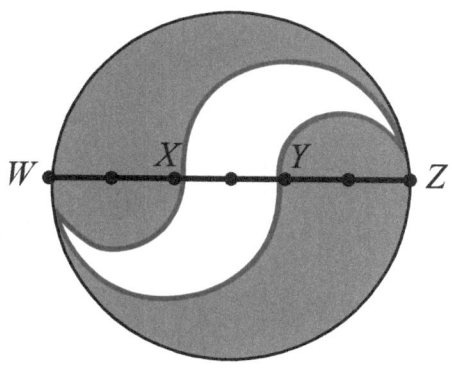

A. 12π

B. 18π

C. 24π

D. 28π

E. 36π

Plane Geometry Mixed Problem Set
Geometry Problem Set 33

56. The figure below consists of four circles, each with its center on segment $\overline{AE}$. Point B is the center of the largest circle. If $\overline{BC} = \overline{CD} = \overline{DE} = 2$, what is the area of the shaded region?

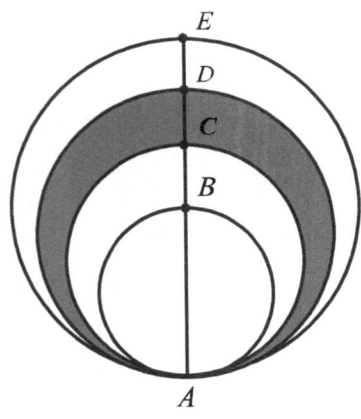

F. 3π
G. 5π
H. 7π
J. 9π
K. 36π

57. In the right $\triangle LMN$ below, $\overline{OP} \parallel \overline{LN}$, and O is the midpoint of $\overline{ML}$. What is the area of the shaded rectangular region?

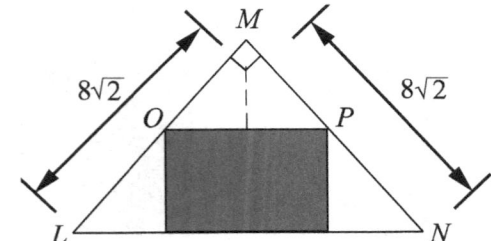

A. 16
B. $16\sqrt{2}$
C. 322
D. 32
E. 64

Plane Geometry Mixed Problem Set
Geometry Problem Set 33

58. If the point O is the center of the circle shown below and the length of the darkened arc AB is 4π and the radius of the circle is 16, what is the probability that a point chosen at random in the circle will fall in the shaded region?

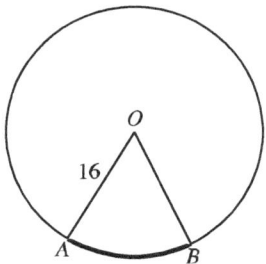

- **F.** $\dfrac{1}{8}$
- **G.** $\dfrac{2}{5}$
- **H.** $\dfrac{1}{6}$
- **J.** $\dfrac{1}{4}$
- **K.** $\dfrac{1}{10}$

59. If the information in the picture below is true and $\overline{RT}$ is perpendicular to $\overline{TS}$, what is the perimeter of $\triangle RTS$?

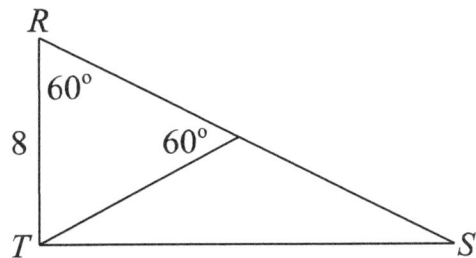

- **A.** $16 + 4\sqrt{3}$
- **B.** $24 + 8\sqrt{2}$
- **C.** 32
- **D.** $16 + 8\sqrt{3}$
- **E.** $24 + 8\sqrt{3}$

Plane Geometry Mixed Problem Set
Geometry Problem Set 33

60. In △ABC below, the measure of ∠A is 30° and the measure of ∠C is 45°. What is the length of segment $\overline{AC}$?

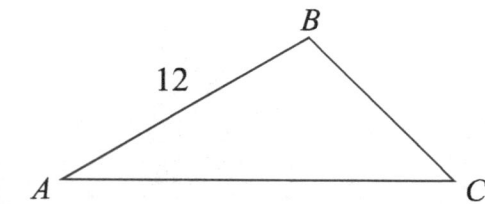

F. 18
G. 24
H. $12\sqrt{2}$
J. $6\sqrt{3}+6$
K. $6\sqrt{2}+6$

DO YOUR FIGURING HERE

61. In the figure below, if the circle with center C has a circumference of 40π, what is the area of △ABC?

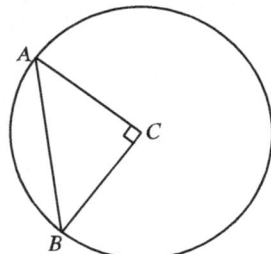

A. 50
B. 100
C. 150
D. 200
E. 400

Plane Geometry Mixed Problem Set
Geometry Problem Set 33

62. In the figure below, $\overline{AX}$ is a radius of the circle with center A. Which of the following triangles has the least area?

- **F.** △RAX
- **G.** △QAX
- **H.** △PAX
- **J.** △SAX
- **K.** △TAX

63. A rectangular park 12 miles long by 8 miles wide is represented on a map with an area of 864 square inches. On the map, how many miles are represented by a length of 1 foot?

- **A.** 2
- **B.** 3
- **C.** 4
- **D.** 6
- **E.** 8

64. △ABC is an equilateral triangle and $\overline{AB}$ is the diameter of a circle with circumference 24π. What is the area of △ABC?

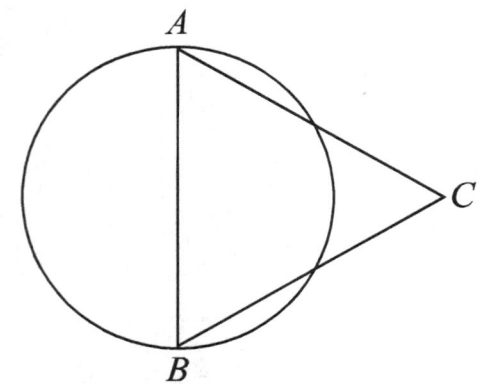

F. 36
G. 72
H. 144
J. $72\sqrt{3}$
K. $144\sqrt{3}$

Plane Geometry Mixed Problem Set
Geometry Problem Set 33

Answer Key

#	Answer	Frequency	Difficulty
1	A	popular	1
2	G	popular	1
3	C	popular	2
4	F	popular	2
5	D	popular	2
6	G	popular	2
7	C	popular	2
8	K	popular	3
9	A	popular	2
10	H	average	1
11	D	popular	1
12	H	popular	2
13	B	popular	2
14	F	popular	2
15	C	popular	2
16	G	popular	2
17	D	popular	2
18	J	popular	2
19	C	rare	3
20	J	rare	3
21	D	popular	3
22	H	popular	2
23	B	popular	2
24	H	popular	3
25	E	popular	3
26	G	popular	4
27	C	average	1
28	J	average	2
29	B	average	4
30	G	popular	1
31	9	popular	2
32	F	average	2
33	A	average	3
34	J	popular	3
35	E	popular	4
36	J	average	1
37	E	average	2
38	F	popular	2
39	C	popular	2
40	K	popular	2
41	C	popular	3
42	J	popular	3
43	D	popular	2
44	12	popular	3
45	B	average	3
46	J	popular	3
47	E	average	3
48	H	popular	4
49	D	popular	4
50	J	popular	4
51	A	popular	2
52	K	popular	3
53	C	popular	2
54	K	popular	3
55	C	average	4
56	J	popular	4
57	D	popular	4
58	F	popular	4
59	E	popular	3
60	J	popular	4
61	D	rare	3
62	H	popular	1
63	C	popular	5
64	K	popular	3

Surface Area
Geometry Problem Set 34

1. If the surface area of a cube is 24, what is the length of one side?

    A. 2
    B. $2\sqrt{2}$
    C. 4
    D. $2\sqrt{6}$
    E. 8

2. What is the surface area of the rectangular prism in the figure?

    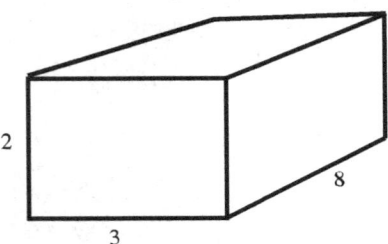

    F. 36
    G. 48
    H. 78
    J. 92
    K. 108

3. Two cubical blocks, each with 3-inch edges, are glued together as shown. What is the surface area, in square inches, of the new solid?

    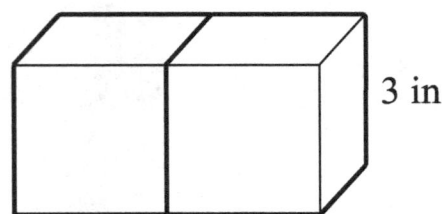

    A. 90
    B. 60
    C. 100
    D. 104
    E. 108

Surface Area
Geometry Problem Set 34

4. Amy is painting the four walls and ceiling of her cube-shaped room. If she has used 8 gallons of paint to cover one wall, how many more gallons of paint does she need?

   F. 32
   G. 28
   H. 24
   J. 16
   K. 8

5. For each face on the cube below, the pattern of shading is the same on the opposite faces. What is the total number of small squares that are shaded on the cube?

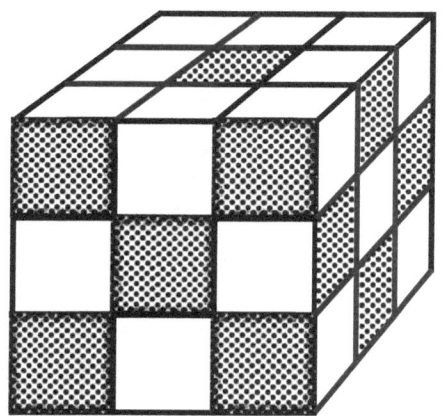

   A. 18
   B. 19
   C. 20
   D. 21
   E. 22

Surface Area
Geometry Problem Set 34

6. If rectangle ABCD, with length 8 and height 4, is rolled into the shape of a cylinder so that there is no overlap, what is the area of one base of the cylinder?

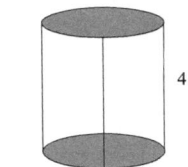

F. $\dfrac{4}{\pi}$

G. $\dfrac{8}{\pi}$

H. $\dfrac{16}{\pi}$

J. $4\pi^2$

K. $8\pi^2$

DO YOUR FIGURING HERE

7. The figure below is known as a triangular prism. Its surface is made up of rectangular and triangular faces, as shown. Each rectangular face has area sl, and each triangular face has area $\dfrac{sh}{2}$. What is the total surface area of the figure, in terms of s, h, and l?

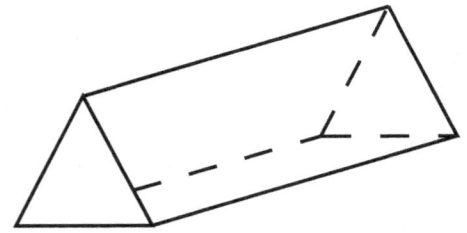

A. $\dfrac{sh}{2} + sl$

B. $sh + sl$

C. $\dfrac{sh}{2} + 3sl$

D. $sh + 3sl$

E. $\dfrac{sh}{2} + 4sl$

Surface Area
Geometry Problem Set 34

8. If a 4-inch cube of cheese were cut in half in all three directions as shown below, then how much greater would the total surface area of the separated smaller cubes be than the surface area of the original 4-inch cube?

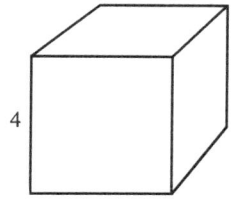
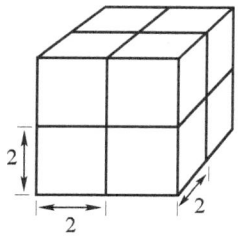

F. 32
G. 48
H. 96
J. 144
K. 192

DO YOUR FIGURING HERE

Surface Area
Geometry Problem Set 34

Answer Key

#	Answer	Frequency	Difficulty
1	A	popular	1
2	J	popular	1
3	A	popular	3
4	F	popular	2
5	C	popular	3
6	H	popular	4
7	D	popular	3
8	H	popular	4

Volume
Geometry Problem Set 35

1. A container in the shape of a right circular cylinder is 12 inches high and has a capacity of 3 quarts. What is the number of quarts of liquid in the container when it is filled to a height of 4 inches?

   A. $\frac{3}{4}$

   B. 1

   C. $1\frac{1}{4}$

   D. $1\frac{1}{2}$

   E. 2

2. John is filling a sandbox for his children. The sandbox is shaped like a rectangular prism, with sides of length 6 feet and 8 feet and a depth of 2 feet. He is buying sand in 10-pound bags that fill 1 cubic foot with sand. How many pounds of sand does John need to fill the sandbox?

   F. 480

   G. 560

   H. 720

   J. 960

   K. 1,280

3. What is the volume, in cubic feet, of a cube with edges of length 5 feet?

   A. 25 cubic feet

   B. 75 cubic feet

   C. 125 cubic feet

   D. 150 cubic feet

   E. 625 cubic feet

Volume
Geometry Problem Set 35

4. The image below shows the dimensions of an ancient structure constructed of identical cubic blocks with side lengths of 2 meters. What is the volume of the structure?

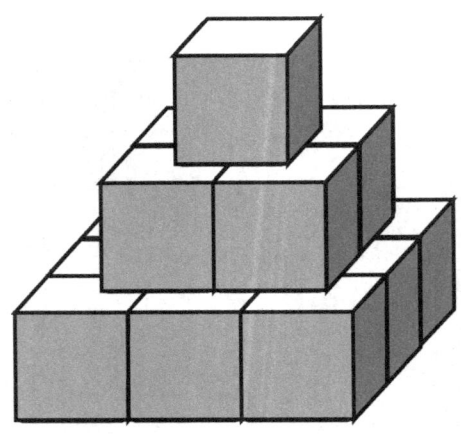

F. 56
G. 72
H. 96
J. 100
K. 112

5. A cube has 2 faces painted black and the remaining faces painted white. The total area of the white faces is 64 square inches. What is the volume of the cube, in cubic inches?

A. 64
B. 125
C. 128
D. 216
E. 256

Volume
Geometry Problem Set 35

6. Each of the glasses below has a radius, r, of 4 inches and a height, h, of 12 inches. The conical glass is filled with water. If the water is emptied completely into the cylindrical glass, what will be the depth in inches of the water in the cylinder?

   (Volume of cone $= \frac{1}{3} \left(\pi r^2 \right) h$)

   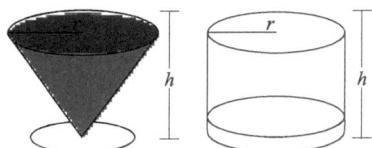

   F. 2
   G. 4
   H. 6
   J. π
   K. 8

7. The circumference of the base of a cylinder is 6π. If the volume of the cylinder is 252π, what is the height?

   A. 7
   B. 24
   C. 28
   D. 36
   E. 42

8. How many cubical blocks, each with edges of length 4 centimeters, are needed to fill a rectangular box that has inside dimensions 20 centimeters by 24 centimeters by 32 centimeters?

   F. 38
   G. 96
   H. 192
   J. 240
   K. 384

Volume
Geometry Problem Set 35

9. What is the length, in inches, of an edge of a cube that has the same volume as a rectangular solid with length 3 inches, width $\frac{3}{4}$ inch, and height $\frac{3}{2}$ inches?

10. The volume of the rectanuglar solid is 648. If $\overline{LO}$ is $2x$, $\overline{OS}$ is $3x$, and $\overline{NO}$ is $4x$, then what is the value of x?

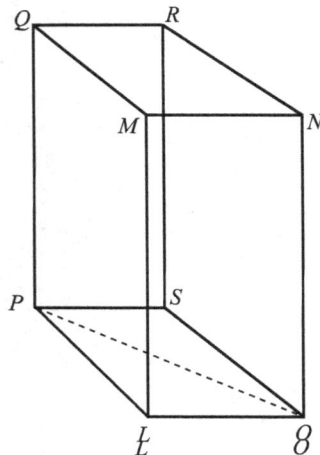

   F. 1
   G. 2
   H. 3
   J. 4
   K. 5

11. The volume of a large fish tank at an aquarium is 702 cubic feet. The tank is 6 feet wide and 13 feet long. How tall is the tank?

   A. 6 feet
   B. 9 feet
   C. 12 feet
   D. 15 feet
   E. 18 feet

Volume
Geometry Problem Set 35

12. In a windowless, cube-shaped room, the ceiling and 4 walls, including the door, are all completely painted. If the floor is not painted, and the painted area is equal to 125 square meters, what is the volume of the room, in cubic meters?

    F. 125
    G. 625
    H. 2,500
    J. 3,125
    K. 15,625

13. Which of the rectangular solids shown below has a volume closest to the volume of a right circular cylinder with radius 1.5 and height 4?

    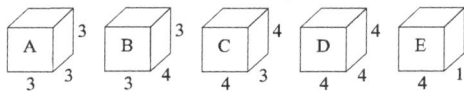

    A. Cube A
    B. Cube B
    C. Cube C
    D. Cube D
    E. Cube E

14. Cube M has volume x. Cube P has edges that are twice as long as Cube M. What is the volume of Cube P, in terms of x?

    F. $2x$
    G. $4x$
    H. $6x$
    J. $8x$
    K. $12x$

Volume
Geometry Problem Set 35

15. Right circular cylindrical candles are sold in rectangular boxes. If x represents the diameter of the candle and y represents the height, in terms of x and y what is the volume of the smallest rectangular box that can be used to sell the candle?

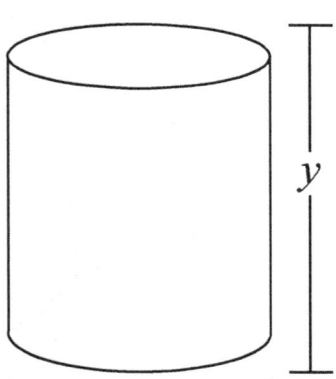

A. $4xy$

B. $2x^{2y}$

C. x^2y

D. $4xy^2$

E. $4x^{2y}$

16. The circumference of the base of a cylinder is 8π. If the volume of the cylinder is 128π, what is the distance of the two furthest points on the cylinder?

F. $4\sqrt{2}$

G. $8\sqrt{2}$

H. $12\sqrt{2}$

J. $16\sqrt{2}$

K. $32\sqrt{2}$

Volume
Geometry Problem Set 35

Answer Key

#	Answer	Frequency	Difficulty
1	B	rare	2
2	J	rare	2
3	C	popular	2
4	K	popular	3
5	A	popular	4
6	G	popular	4
7	C	popular	3
8	J	popular	3
9	$\frac{3}{2}$	popular	3
10	H	popular	2
11	B	popular	3
12	F	popular	3
13	A	popular	3
14	J	popular	3
15	C	popular	4
16	G	popular	5

Diagonals
Geometry Problem Set 36

1. The pencil-holder in the figure below is a cylinder that is 9 inches high and has a radius of 4 inches. If there are 5 pencils having lengths of 7 inches, 10.5 inches, 12 inches, 12.5 inches, and 14 inches, how many CANNOT fit entirely inside the can?

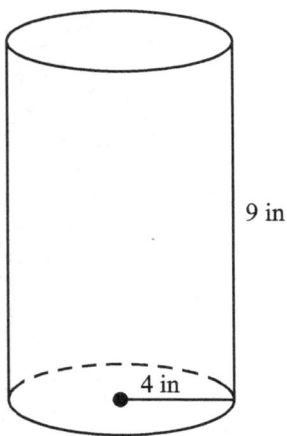

A. 0
B. 1
C. 2
D. 3
E. 4

Diagonals
Geometry Problem Set 36

2. The cube shown below has edges that each measure 4. Points A and B are the midpoints of the two edges. What is the length of $\overline{AB}$ (not shown)?

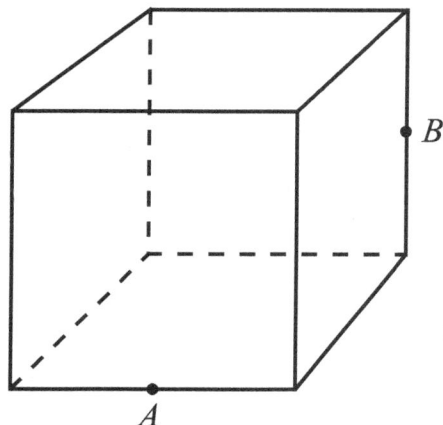

F. 2
G. 4
H. $2\sqrt{6}$
J. $4\sqrt{2}$
K. $6\sqrt{2}$

Diagonals
Geometry Problem Set 36

3. The cube below has edges that each measure 4 and x and y are the midpoints of their respective edges. What is the length of line segment $\overline{xy}$ (not shown)?

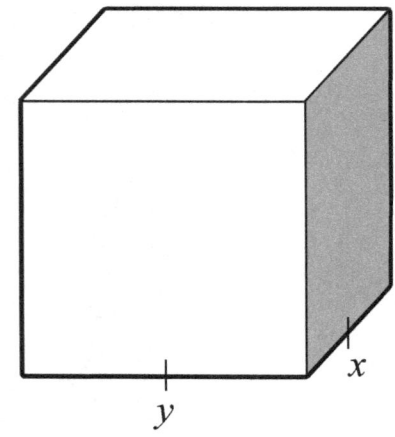

A. $\sqrt{6}$

B. $2\sqrt{2}$

C. $\sqrt{22}$

D. $2\sqrt{6}$

E. $4\sqrt{2}$

4. The height of a right circular cylinder is 5 and the diameter of its base is 4. What is the distance from the center of one base to a point on the circumference of the other base?

F. 3

G. 5

H. $\sqrt{29}$ (approximately 5.39)

J. $\sqrt{33}$ (approximately 5.74)

K. $\sqrt{41}$ (approximately 6.40)

Diagonals
Geometry Problem Set 36

5. The pyramid $ABCDE$ pictured below has an altitude h and a square base with side lengths of s. Point E is located at the vertex, and each vertical edge equals v. If $v = s$, which of the following is equal to the length of $\overline{AC}$, in terms of h?

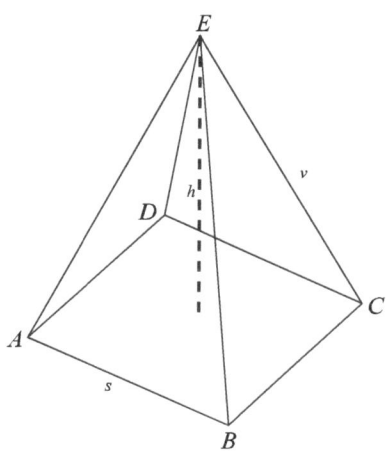

   A. h
   B. $2h$
   C. $4h$
   D. $\dfrac{(h\sqrt{2})}{2}$
   E. $h\sqrt{2}$

6. A cube with a volume of 27 cubic inches is inscribed in a sphere so that each vertex of the cube touches the edge of the sphere. What is the length of the diameter, in inches, of the sphere?

   F. 3
   G. $2\sqrt{3}$
   H. $3\sqrt{2}$
   J. $3\sqrt{3}$
   K. 9

Diagonals
Geometry Problem Set 36

Answer Key

#	Answer	Frequency	Difficulty
1	C	rare	2
2	H	rare	2
3	B	rare	1
4	H	rare	2
5	B	rare	5
6	J	rare	2

Ratios and Dimensions
Geometry Problem Set 37

1. How many square feet are in 9 square yards?
   - A. 3 ft^2
   - B. 9 ft^2
   - C. 27 ft^2
   - D. 63 ft^2
   - E. 81 ft^2

2. If the radius and height of a cylinder are both tripled, how many times greater is the volume of the new cylinder than the original cylinder?
   - F. 6
   - G. 9
   - H. 12
   - J. 27
   - K. 36

Ratios and Dimensions
Geometry Problem Set 37

Answer Key

#	Answer	Frequency	Difficulty
1	E	rare	3
2	J	rare	3

Logical Spatial Relationships
Geometry Problem Set 38

1. What is the total number of right angles formed by the edges of a cube?

    A. 12
    B. 16
    C. 20
    D. 24
    E. 36

2. If the rectangular box with no lid shown below is cut along the vertical edges and flattened, which of the following figures best represents the result?

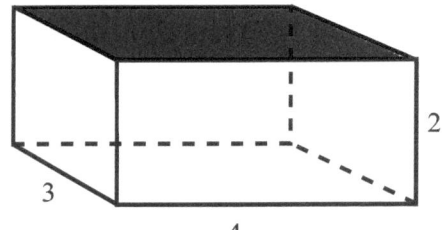

F.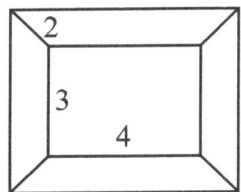

G.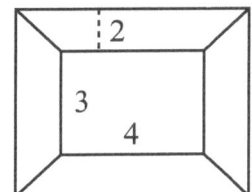

H.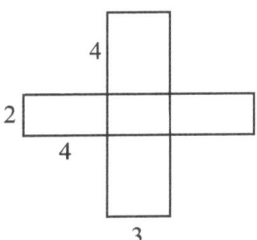

J.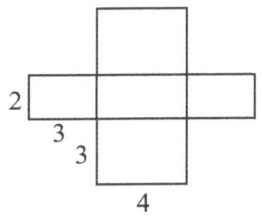

K.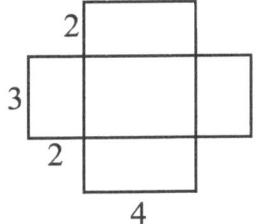

DO YOUR FIGURING HERE

Logical Spatial Relationships
Geometry Problem Set 38

3. In the cube below points A, C, and E are midpoints of their respective edges. Which of the following is the largest angle?

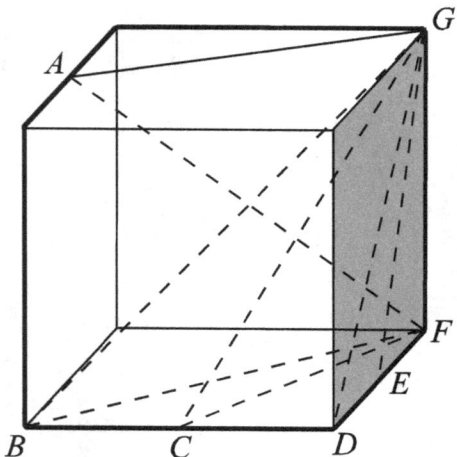

A. $\angle GAF$

B. $\angle GBF$

C. $\angle GCF$

D. $\angle GDF$

E. $\angle GEF$

4. What is the volume, in cubic inches, of the smallest rectangular box into which a 1-inch cube, a 2-inch cube and a 3-inch cube can be packed together? (Assume that the cubes are solid.)

F. 36

G. 45

H. 54

J. 75

K. 125

Logical Spatial Relationships
Geometry Problem Set 38

5. A large solid cube is made by gluing together a certain number of smaller, unpainted cubes. After the smaller cubes are glued together, the six faces of the large cube are decorated with paint. If exactly 64 of the small cubic blocks have no blue paint on them, how many small cubic blocks make up the larger cube?

   A. 48
   B. 96
   C. 116
   D. 216
   E. 324

6. A large solid cube is made by gluing together a certain number of identical white cubes. After the smaller cubes are glued together, the six faces of the large cube are painted blue. If exactly 27 of the small cubic blocks have no blue paint on them, how many small cubic blocks make up the larger cube?

   F. 25
   G. 75
   H. 100
   J. 125
   K. 175

7. Two spheres, one with radius 7 and one with radius 4, are tangent to each other. If P is any point on one sphere and Q is any point on the other sphere, what is the maximum possible length of $\overline{PQ}$?

   A. 7
   B. 11
   C. 14
   D. 18
   E. 22

Logical Spatial Relationships
Geometry Problem Set 38

8. The three-dimensional figure below has two parallel bases and 15 edges. Line segments are to be drawn connecting vertex V with each of the other 9 vertices in the figure. How many of the segments will NOT lie on an edge of the figure?

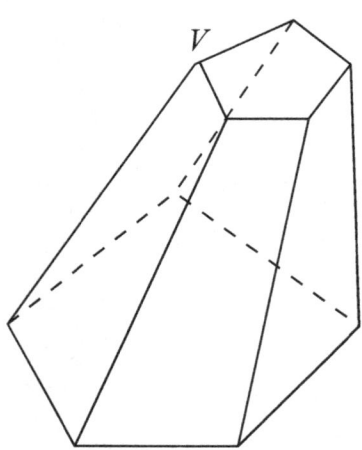

9. The figure below shows a solid with a hexagonal base. If each edge of the base has length 5 and each of the other edges of the solid has length 15, what is the sum of the lengths of all 12 edges?

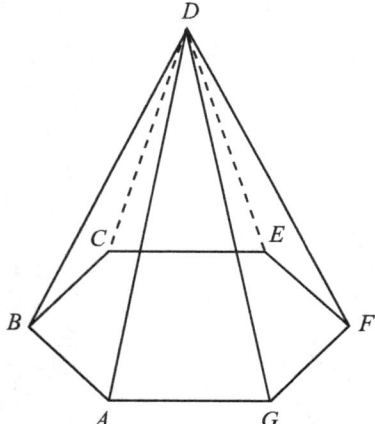

A. 100
B. 105
C. 110
D. 115
E. 120

Logical Spatial Relationships
Geometry Problem Set 38

10. The surface of a 3-dimensional solid consists of faces, each of which has the shape of a polygon. What is the least number of such faces that the solid can have?

    F. 2
    G. 3
    H. 4
    J. 5
    K. 6

11. A rectangular prism has faces with areas of 15, 21, and 35 square inches. What is the volume, in cubic inches, of the prism?

    A. 105
    B. 315
    C. 1,250
    D. 5,512.5
    E. 11,025

Logical Spatial Relationships
Geometry Problem Set 38

Answer Key

#	Answer	Frequency	Difficulty
1	D	rare	1
2	K	rare	1
3	E	rare	4
4	G	rare	3
5	D	rare	1
6	J	rare	4
7	E	rare	1
8	6	rare	2
9	E	rare	1
10	H	rare	2
11	A	rare	3

Intro to Trig
Geometry Problem Set 39

1. If $b = 4a$, then what is $\tan(\angle ABC)$?

   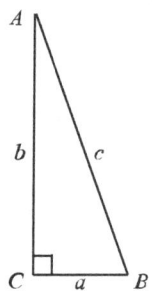

   A. $\dfrac{1}{2}$

   B. $\dfrac{1}{\sqrt{2}}$

   C. 2

   D. $2\sqrt{2}$

   E. 4

2. For right triangle $\triangle RST$, what is $\cos R$?

   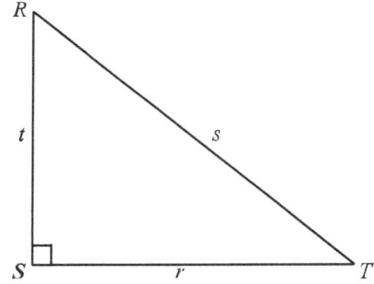

   F. $\dfrac{r}{s}$

   G. $\dfrac{r}{t}$

   H. $\dfrac{t}{r}$

   J. $\dfrac{t}{s}$

   K. $\dfrac{s}{t}$

Intro to Trig
Geometry Problem Set 39

3. The hypotenuse of the right triangle △KLM shown below is 12 feet long. The sine of ∠M is $\frac{3}{5}$. About how many feet long is $\overline{KL}$?

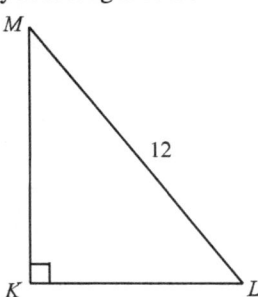

- A. 6.0
- B. 7.2
- C. 8.4
- D. 14.3
- E. 20.0

4. In the right triangle shown below, which of the following statements is true about ∠A?

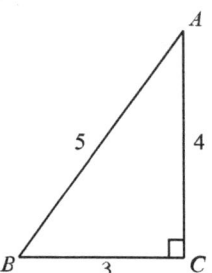

- F. $\cos A = \frac{4}{5}$
- G. $\sin A = \frac{4}{5}$
- H. $\tan A = \frac{4}{5}$
- J. $\cos A = \frac{5}{4}$
- K. $\sin A = \frac{5}{4}$

Intro to Trig
Geometry Problem Set 39

5. The lengths, in feet, of the sides of right triangle $\triangle ABC$ are as shown in the diagram below. What is the cotangent of $\angle A$, in terms of a?

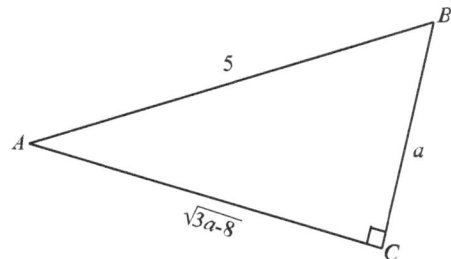

A. $\sqrt{3a - 8}$

B. $\dfrac{5}{a}$

C. $\dfrac{a}{5}$

D. $\dfrac{a}{\sqrt{3a - 8}}$

E. $\dfrac{\sqrt{3a - 8}}{a}$

6. For an angle with measure α in a right triangle, $\cos \alpha = \dfrac{4}{5}$ and $\tan \alpha = \dfrac{3}{4}$. What is the value of $\sin \alpha$?

F. $\dfrac{3}{4}$

G. $\dfrac{3}{5}$

H. $\dfrac{4}{5}$

J. $\dfrac{5}{3}$

K. $\dfrac{5}{4}$

Intro to Trig
Geometry Problem Set 39

7. A right triangle $\triangle ABC$ is shown below. For any such triangle, $(\tan A)(\cos B)$ is equivalent to:

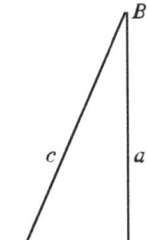

A. $\dfrac{c}{a}$

B. $\dfrac{b^2}{ac}$

C. $\dfrac{a^2}{bc}$

D. $\dfrac{ab}{c^2}$

E. $\dfrac{a}{c}$

DO YOUR FIGURING HERE

8. A dog is tied to a 20-foot leash, which is at an angle of $30°$ to the ground. Which of the following expressions gives the distance, in feet, from the dog's collar to point A?

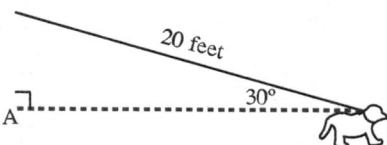

F. $20 \cos 30°$

G. $20 \sin 30°$

H. $20 \tan 30°$

J. $\dfrac{20}{\cos 30°}$

K. $\dfrac{20}{\sin 30°}$

Intro to Trig
Geometry Problem Set 39

9. The diagram of the roof for a new storage shed is shown below. Lengths are given in yards. Which expression gives the length in meters of the vertical support, $\overline{BD}$?

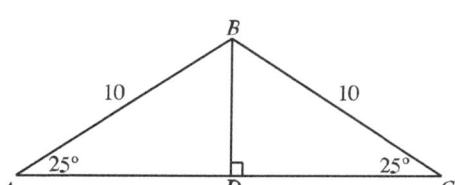

A. $10 \sin 25°$

B. $10 \tan 30°$

C. $10 \cos 30°$

D. $\dfrac{10}{\cos 25°}$

E. $\dfrac{10}{\sin 25°}$

Intro to Trig
Geometry Problem Set 39

10. A ladder with a length of $\sqrt{73}$ feet is placed 3 feet from the base of a house. It touches the house 8 feet above the ground. Which of the following expressions represents the measure of the angle that the ladder will make with the level ground?

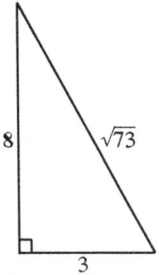

- **F.** $\tan^{-1}\left(\dfrac{3}{\sqrt{73}}\right)$
- **G.** $\tan^{-1}\left(\dfrac{8}{3}\right)$
- **H.** $\tan^{-1}\left(\dfrac{\sqrt{73}}{8}\right)$
- **J.** $\tan^{-1}\left(\dfrac{3}{8}\right)$
- **K.** $\tan^{-1}\left(\dfrac{8}{\sqrt{73}}\right)$

11. For an angle with measure α in a right triangle, $\sin\alpha = \dfrac{10}{\sqrt{164}}$ and $\tan\alpha = \dfrac{10}{8}$. What is the value of $\cos\alpha$?

- **A.** $\dfrac{8}{\sqrt{164}}$
- **B.** $\dfrac{10}{\sqrt{164}}$
- **C.** $\dfrac{4}{5}$
- **D.** $\dfrac{5}{4}$
- **E.** $\dfrac{\sqrt{164}}{8}$

Intro to Trig
Geometry Problem Set 39

12. The figure below shows a person looking down at a car on the ground level through a pair of binoculars. The binoculars are 60 feet above the ground and there is an angle of depression of 53° to the car. What is the horizontal distance between the binoculars and the car? Round to the nearest tenth of a foot.
(Note: $\sin 53° \approx 0.80$, $\cos 53° \approx 0.60$, $\tan 53° \approx 1.33$)

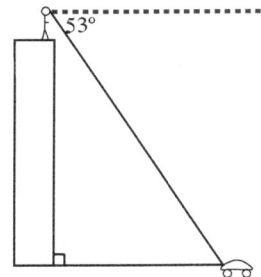

F. 36.0

G. 45.1

H. 58.2

J. 75.0

K. 100.0

DO YOUR FIGURING HERE

Intro to Trig
Geometry Problem Set 39

13. Sarah is loading a truck using a ramp, as shown below. The ramp is 9 feet long, and the end of the ramp that is resting on the truck is 2 feet above the level ground. Which of the following expressions gives the angle of inclination of the ramp?

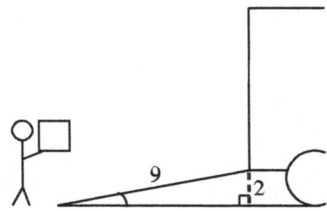

A. $\arccos\left(\frac{\sqrt{85}}{2.5}\right)$

B. $\arctan\left(\frac{\sqrt{85}}{9}\right)$

C. $\arcsin\left(\frac{2}{9}\right)$

D. $\arccos\left(\frac{2}{9}\right)$

E. $\arcsin\left(\frac{9}{2}\right)$

Intro to Trig
Geometry Problem Set 39

14. The radius of a right circular cone shown below is 6 inches and the height of the cone is 9 inches. Solving which of the following equations gives the measure, θ, of the angle formed by a slant height of the cone and a radius?

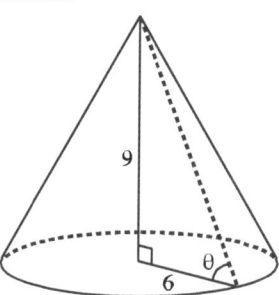

F. $\sin\theta = \dfrac{3}{2}$

G. $\sin\theta = \dfrac{2}{3}$

H. $\cos\theta = \dfrac{3}{2}$

J. $\tan\theta = \dfrac{3}{2}$

K. $\tan\theta = \dfrac{2}{3}$

15. The side lengths of right triangle $\triangle QRS$ are given in inches in the figure below. What is $\tan Q$?

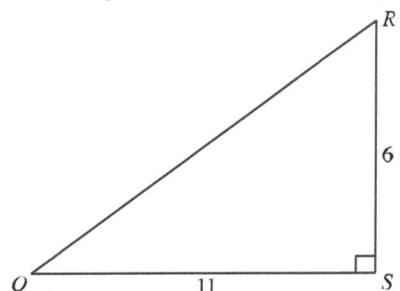

A. $\dfrac{6}{11}$

B. $\dfrac{6}{\sqrt{157}}$

C. $\dfrac{11}{6}$

D. $\dfrac{11}{\sqrt{157}}$

E. $\dfrac{\sqrt{157}}{11}$

Intro to Trig
Geometry Problem Set 39

16. The figure below shows a 12-foot ladder leaning against a vertical wall. The base of the ladder makes a 60° angle with the level ground. Which of the following expressions gives the distance, in feet, between the ground and the point where the ladder meets the wall?

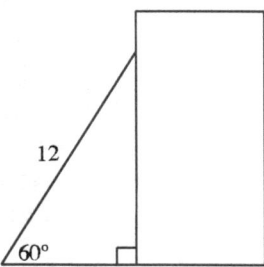

- **F.** $12 \sin 60°$
- **G.** $12 \cos 60°$
- **H.** $12 \tan 60°$
- **J.** $\dfrac{12}{\sin 60°}$
- **K.** $\dfrac{12}{\sin 60°}$

17. In the figure below, 16-foot ramp forms an angle of 14° with the ground, which is horizontal. Which of the following is an expression for the height, in feet, of the ramp?

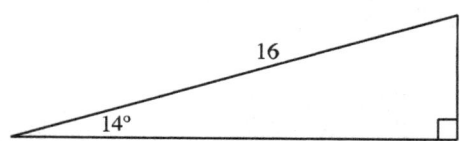

- **A.** $16 \sec 14°$
- **B.** $16 \cot 14°$
- **C.** $16 \tan 14°$
- **D.** $16 \cos 14°$
- **E.** $16 \sin 14°$

Intro to Trig
Geometry Problem Set 39

18. A movie projector emits light as shown in the top-view diagram below. To make the light projection cover the full width of a 12-foot screen, the projector must be placed a certain distance from the screen. Which of the following expressions gives the distance, in feet, from the projector to the screen?

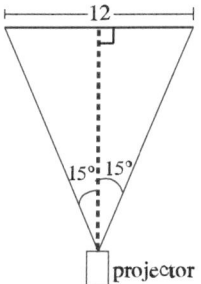

F. $6 \sec 15°$

G. $6 \cot 15°$

H. $6 \tan 15°$

J. $6 \cos 15°$

K. $6 \sin 15°$

19. In the figure below, $\tan \theta =$

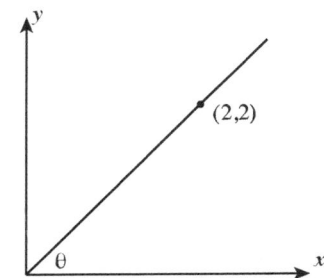

A. $\dfrac{1}{2}$

B. $\dfrac{\sqrt{2}}{2}$

C. $\dfrac{2}{\sqrt{2}}$

D. 1

E. $\sqrt{2}$

Intro to Trig
Geometry Problem Set 39

20. For the right triangle shown below, with the dimensions given in centimeters, which of the following has a value of $\frac{4}{5}$?

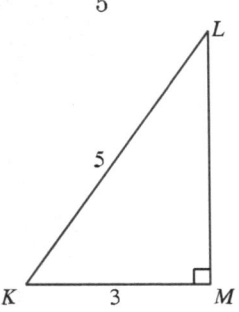

- **F.** $\sin K$
- **G.** $\sin L$
- **H.** $\cos K$
- **J.** $\tan K$
- **K.** $\tan L$

21. An obelisk, shown below, casts a shadow that is 100 yards long. The angle of elevation, θ, from the tip of the shadow to the top of the building has a sine of $\frac{4}{5}$. What is the height of the obelisk in yards? Round to the nearest tenth of a yard.

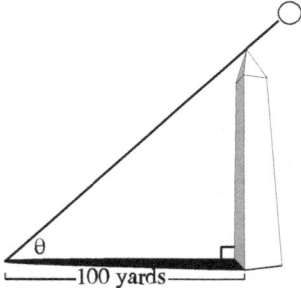

- **A.** 98.2
- **B.** 113.8
- **C.** 125.0
- **D.** 133.3
- **E.** 166.7

Intro to Trig
Geometry Problem Set 39

Answer Key

#	Answer	Frequency	Difficulty
1	E	popular	1
2	J	popular	1
3	B	popular	2
4	F	popular	1
5	E	popular	3
6	G	popular	2
7	C	popular	3
8	F	popular	1
9	A	popular	2
10	G	popular	1
11	A	popular	2
12	G	popular	2
13	C	popular	1
14	J	popular	1
15	A	popular	1
16	F	popular	1
17	E	popular	1
18	G	popular	2
19	D	popular	2
20	F	popular	3
21	D	popular	3

Formulas
Quick Drill

1. What is the area of a rectangle?

2. What is the area of a square?

3. What is the volume of a rectangular prism?

4. What is the volume of a cube?

5. What is the surface area of a rectangular prism?

6. What is the surface area of a cube?

7. What is the area of a circle?

8. What is the circumference of a circle?

9. What is the volume of a cylinder?

10. What is the surface area of a cylinder?

11. What is the distance formula?

12. What is the 3-dimensional distance formula?

13. What is the midpoint formula?

14. What is the equation of a line in slope-intercept form?

15. What does m stand for and what is the formula for it?

16. What does b stand for?

17. What is the area of a triangle?

18. What is the formula for finding the hypotenuse of a right triangle? What is it called?

19. Label all sides in the 30-60-90 triangle.

Formulas
Quick Drill

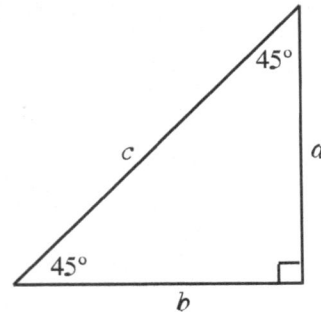

20. Label all sides in the 45-45-90 triangle.

21. What is the area of a parallelogram?

22. What is the area of a trapezoid?

23. What is the sum of the measure of the interior angles of a polygon?

24. What is the measure of one interior angle of a regular polygon?

25. What is the quadratic formula?

26. What is the formula for arithmetic mean?

27. What is a median?

28. What is a mode?

29. What is the formula for the common difference of an arithmetic sequence?

30. What is the formula for the nth term of an arithmetic sequence?

31. What is the formula for the sum of n terms of an arithmetic sequence?

32. What is the formula for the common ratio of a geometric sequence?

33. What is the formula for translating radians into degrees?

34. What is the equation for a circle?

35. What is the equation for an ellipse?

36. What does SOHCAHTOA stand for?

Formulas
Quick Drill
Answer Key

#	Answer
1	$A = lw$
2	$A = s^2$
3	lwh
4	$V = s^3$
5	$2(lw + hl + hw)$
6	$SA = 6s^2$
7	$A = \pi r^2$
8	$C = \pi d$
9	$V = \pi r^2 h$
10	$SA = 2\pi r^2 + \pi dh$
11	distance = $\sqrt{(x_2 - x_1)^2 + (y_2 - y_1)^2}$
12	distance = $\sqrt{(x_2 - x_1)^2 + (y_2 - y_1)^2 + (z_2 - z_1)^2}$
13	midpoint = $\left(\dfrac{x_1 + x_2}{2}, \dfrac{y_1 + y_2}{2}\right)$
14	$y = mx + b$
15	slope = $\dfrac{y_2 - y_1}{x_2 - x_1}$
16	y-intercept
17	$A = \dfrac{bh}{2}$
18	$a^2 + b^2 = c^2$, Pythagorean Theorem
19	$a = x$, $b = x\sqrt{3}$, $c = 2x$
20	$a = x$, $b = x$, $c = x\sqrt{2}$
21	$A = bh$
22	$A = \dfrac{h(b_1 + b_2)}{2}$
23	$180(n - 2)$
24	$\dfrac{180(n - 2)}{n}$
25	$x = \dfrac{-b \pm \sqrt{b^2 - 4ac}}{2a}$
26	$\dfrac{1}{n}\sum_{i=1}^{n} x_i$
27	the middle number in a set numerical order
28	the most common number in a set
29	$d = a_n - a_{n-1}$
30	$a_n = a_1 + (n - 1)d$
31	$\sum_{i=1}^{n} a_i = \dfrac{n}{2}(a_1 + a_n)$
32	$r = \dfrac{a_n}{a_{n-1}}$
33	degree = radian $\left(\dfrac{180}{\pi}\right)$
34	$(x - h)^2 + (y - k)^2 = r^2$, where (h, k) is the center
35	$\dfrac{(x - h)^2}{a^2} + \dfrac{(y - k)^2}{b}$ where (h, k) is the center and $2a$ and $2b$ are the lengths of the axes
36	sine = $\dfrac{\text{opposite}}{\text{hypotenuse}}$, cosine = $\dfrac{\text{adjacent}}{\text{hypotenuse}}$, and tangent = $\dfrac{\text{opposite}}{\text{adjacent}}$